EVERYTHING YOU NEED TO KNOW ABOUT PHYSICS &… EVERYTHING!

Bongani Banda

For permission requests, write to:
Bongani Banda, bandabongani645@gmail.com

First Edition

To my family and friends, who inspire me and care for me,
To dream big, continue further.

Table of Contents

Preface

I wrote this book to inspire other readers out there, who are career-driven and curious for physics, and who are there to dream bigger and continue a journey in life. The book teaches everything you'd like to know about Physics and its relation to every matter we have in this world. Like time, space, gravity, the universe and gaming. The book that I have written is a journey from the beginning of the universe to the future of physics.

I, the author, Bongani Banda, loved physics since I was a young kid, when I was watching a science educational show on Da Vinci Kids, and started planning crazy experiments. This book teaches why physics or any other field of study helps us to change the world and the universe. I am inspired by renowned physicists like Brian Cox, Jim Al-Khalili, Matt Parker, Hakeem Oluseyi, Stephen Hawking, Neil DeGrasse Tyson and Carl Sagan. I hope, you get inspired and follow your curiosity with the help of this book and know everything about physics and its relations of it. Thanks to all my inspirations who led me to write this book.

From the beginning to the end, from the past to the future, you will learn a lot about Physics and everything about it in this fun-driven book. I hope this book inspires you and your dreams and goals you need to achieve, whether you love Physics or not.

Introduction

Everything You Need To Know About Physics &... Everything

This book focuses on the fundamental world of physics and its relations to the real-world. This book has many answers to your questions like "What is the Big Bang and how did it create the universe?", and "How is physics related to gaming, and how will it affect the real world?". "Everything You Need To Know About Physics &... Everything" is a fun-driven book about the fundamental principles and creations that physics has done for us. From astronomy to the rivalry between metric and imperial system or the battle of matter and antimatter, this is the book that lead you out-of-this-world to outer space, then back in time to the Big Bang, causing the creation of our universe.

This book is educational and explores the main concepts of physics, from SI units to the laws of motion. This book also shows ideas of what physicists will have to take a next step towards using physics into the near and far future. Physics is not just for physicists or university students, but it's for everyone to learn and continue to dream big with it and other fields of study.

Like Albert Einstein, Ernest Rutherford, Niels Bohr, Glenn Seaborg and Carl Sagan, and other stupendous scientists out there today, were also inspired by one of their inspiring scientists of the past. I hope the book helps you open your imagination and enter a world full of physics and following your wonder.

CHAPTER 1

EVERYTHING YOU NEED TO KNOW ABOUT PHYSICS & EVERYDAY LIFE

Have you ever wondered about physics resulting in our everyday life? There's true physics behind two cars colliding between each other causing a tragic car accident, which is an example of collision physics. Or two people bumping into each other without seeing themselves approaching on the way. But physics has told us many stories and true factual stories about our world, our universe and beyond.

Physics' connection with our everyday life is obvious like the first-ever creation of our universe from the Big Bang, light in our homes, and we use fire to heat our food during blackouts outside our homes. There is physics in our everyday life, it helps us to do everything, but it might cause very bad things in our daily lives, like in history, the fatalities that took place in Hiroshima in 1945, and Boksburg, South Africa in late 2022.

Ancient Greeks believed that the god of lightning, Zeus made lightning, but it is actually physics that made lightning made of electrons together into the clouds, and attracts with protons on the ground, causing a huge spark. But weather or should I say "Atmospheric physics"? Atmospheric physics has been impacting our everyday life, rain ruining days for young children, or a tornado ruining houses in some parts of the world. Physics might be bad in our lives, but can also be helpful in our daily lives.

As you see in the image above, clouds in the skies cause a very sparking atmospheric phenomena, that people hate and some love. Lightning is an atmospheric phenomena, that any meteorologist and physicist can understand and investigate. It is a force of nature that anyone cannot understand because other people think lightning comes from Zeus from Greek mythology, or Thor from Norse (Scandinavian) mythology, or from our Father God.

Atmospheric physics is a type of physical science that deals with types of weather, like tornadoes, monsoons, rain, hail and more. These atmospheric phenomenons are an example of physics' connection with our everyday life. Even cars with high speed showing velocity in our everyday life, or apples falling showing gravity in our everyday life like Newton perceived the idea of renaming the force "gravita" in Old English. Or explosions showing big chemical reactions in our everyday life, showing a violent and dangerous way of testing how physics is connectedwith our everyday life. Even gaming can help build physics and mathematics skills. Physics can help enrich our daily lives.

CHAPTER 1.1

PHYSICS & GAMING

HOW THESE TWO COLLIDE INTO ONE FASCINATION!

Have you ever wondered how physics and gaming are related to each other? Is gaming bad for our lives? Well, physics is also the gateway to the future of high-quality gaming.

Why is physics the gateway to high-quality gaming? Well, physics is the gateway because modern video games use simulation devices that help video game developers to make the game look realistic and graphical, those devices are called physics engines that help developers to make the video game look realistic like EA Sports FC, developers have been using a physics engine known as Frostbite, which is used only on EA companies around the world to develop a video game like FC. Physics are also used in code and programming when developing a video game on Java, C++ or Python. Gamers around the world don't know that video games are realistic thanks to physics by the help of physics engines that use a way of making the game look realistic called "Realism".

Realism is a concept where video game developers, artists who make art assets for video games and future video games that are developing now. Physics engines like NVIDIA PhysX, Frostbite or EA Sports Ignite to calculate the distance of the game's background or measurement of characters. Physics and gaming collide into one amazing thing, that makes characters interact with objects, that's the true physics' connection with gaming and modern technology. Gaming is not bad for our lives, it helps engage with the gaming world and learn entering a new world like the futuristic devices of gaming, like the Meta Quest VR.

From the past, gaming has start influencing people. From the first console invention of Prof. Ralph Baer to the innovation of PlayStation, without physics, gaming would have not existed based on the movement of characters, what they do, interactions between characters and objects. Physicists have discovered the connection between physics and gaming. Gamers would thank physics for what it has done for them, to play thousands of video games from consoles, PCs or mobile. Physics and gaming have been along since the beginning of the First Computer Era. Between the 1970s and 1980s, the first video game console ever built, the Magnavox Odyssey, was born, filled with amazing 8-bit video games like Tennis for Two, invented by William Higinbotham.

Ralph Baer was a genius to build a device that would connect physics with gaming in the future. Did he know that his first ever console device would make history? Then after, it came the Atari, a Japanese-made video game console, that was built in the mid-1970s and also the invention of Pong and Pac-Man. Today, physics and gaming are being used as one always, and will be used forever until artificial intelligence takes over the world, I think. During video game development, programmers use collision physics to make the character look like it's colliding with another object.

Collision physics is the way of showing two objects colliding together showing a terrible impact. The animation of the video game characters impress the gamers a lot, based on their motion and battle movement based on the velocity and interaction the game is using from developers. For console developers, they have been using SDK stands for Software Development Kit from PlayStation, Nintendo or Xbox to connect ports on PCs or TVs to start developing the game from ideas, to storylines, to character design, and using physics engines to make the game look realistic based on computer physics about resolution on 4K or 8K games. Games in the past were 8-bit resolution because they didn't have physics engines back then. The resolution is calculated by the help of adding the game's image pixels by 180p. The resolution start from 180p multiply by 2 gives 360p multiply by 2 gives 720p by 2 also gives 1080p which is Full HD and so on. Developers use resolution physics to calculate the modern look nd realism of high-quality gaming images when developing a video game. Like this image of these cyber-like soldiers in this modern physics-realistic background below:

Gaming is also bad for our mental health as well, if you play games for more than 3 hours of gameplay, you get addicted and can impact your health. But physics and gaming have a unique chemistry that shapes our modern tech-look of playing video games realistically in the future.

Vectors in gaming, during development, game developers use video game engines that has physics software to calculate the 3D vector dimensions of how video game characters are going to move during animation timelines in each axes. X-Axis, length, Y-Axis, width, and Z-Axis, height. The character stays at the point where its going to be animated. Shows how each character moves in each dimension when animating a game storyline on the previous page.

There's a lot more to embrace the life of gaming by reliving it using virtual reality. Virtual reality is a gaming technology that simulates the game into an open world or a real-world to think that the person is in the game. Physics' connection with gaming is also simulation from sports like football, basketball, golf, wrestling, racing and not just sports, but life.

Gamers think that gaming is just a fun activity, but it is also an activity that helps them learn the new ways of life and explore new worlds in a video game, and Gaming helps you develop maths and physics skills like Portal 2, a video game that focuses on Chell, a silent protagonist who enters Aperture Science, a lab where she will learn to solve physics puzzles. From graphics, motion to collision physics, Physics has a strong connection with gaming. Physics is the gateway to future gaming, and physics is also the importance in game development and video game designs.

CHAPTER 1.2

PHYSICS & THE WORLD WE LIVE IN

TIME, SPACE AND THE UNIVERSE

Have you ever wondered about the world we live in is all surrounded by the fundamental world of physics? But physics has shaped our daily lives in this world we live in. But let's go back in time from the Big Bang, well, our universe was only filled with immense clouds made of hydrogen and helium until a big explosion took place, 13,8 billion years ago.

During the Big Bang, there was no hours, minutes or seconds but there was Planck time which was first demonstrated by German physicist Max Planck in the 20th century. What is a Planck time? Planck time is a unit of time where some events take place faster than a second, or a millisecond or even a zeptosecond, it is the fastest time unit possible. What people don't understand about time, is our Earth's rotation around the Sun. The Earth ends its revolution after 365¼ days because the year is divided into four seasons and the Moon ends its revolution around the Earth for a month or 29½ days.

The Earth rotates around the Sun for a day, meaning it has 365 rotations in one year, or 366 in a leap year. The physics behind time, is that it is also the temporal rate quantity of speed like velocity or acceleration like explained in derivatives in the field of mathematics, for example in velocity or distance, kilometres per hour, and in acceleration, metres per second, to show the distance of how the different modes of transportation travel in every unit of time, as it shows in physics.

People say the earth is the biggest clock in the universe, well, it's true, it is the biggest clock, because the earth has different variants of time zones starting from Greenwich, England. GMT stands for Greenwich Meridian Time, it is the most known time zone in the world. In South Africa, our time zone is GMT+2:00, why? For example, let's say, England's time is 08:00 AM GMT, then in South Africa, what is the time? The time is actually 10:00 AM GMT, why?

Because South Africa's time is 2 hours ahead from England's time. Even time is different beyond our biggest clock. Earth's revolution around the Sun is a year. But further from Earth, Neptune's revolution around the Sun is actually 164 years that's like a year in Neptune but on Earth is more than that. Why on Neptune the revolution takes 164 years? The Neptune's rotation around the Sun is slower than other planets near the Sun like Mercury, Venus and our home, Earth.

Time has also been around in space. Physics' connection with space and time is based on Einstein's special theory of relativity and there is an invisible phenomenon that Albert Einstein has been discovering in the 20th century, known as the spacetime curvature.

The spacetime curvature has been observed by Albert Einstein in his special theory of relativity, he found out that the curvature has three dimensions of space, which are height and length also shown on mathematics when learning 3D shapes and area, and one dimension of time that shows when combined, create the four-dimensional mathematical model that shows the Earth or other planets like Mercury have a huge amount of mass that create a curvature underneath their planetary size, even the Sun, has 99.85% mass of the solar system. Space and time has shaped our world and our universe, from the Big Bang to supernovas. Gravity is the force that surrounds our universe, and it is measured in Newtons, unit of force named after English physicist and theologian, Isaac Newton. Even gravity defies between space and time. What is the connection between spacetime and our universe? People ask, who or what created the Universe? The Universe was created by the Big Bang, a huge violent explosion that occurred 13,8 billion years ago. What made the Big Bang and what was the cause of it? Well, before the Big Bang, before our universe became our universe, space was filled with nothing but hydrogen gas around darkness and reacted with helium gas as it came out of nowhere. Both hydrogen and helium gas violently reacted each other and created the Big Bang!

Since the Big Bang, the universe before was made of heat, at that time when the universe was created, the explosive reaction caused the universe to be full of heat, and a few Planck times later, the universe started expanding and inflation started to cool down. Today, the universe is still expanding, but on recent newspapers, journalists believe that the universe stopped expanding because of climate change shown on Earth. But will this pseudoscience continue to spread? Actually, these newspaper facts spread pseudo-belief on newspaper readers today. But the amazing fact is that, the universe is still expanding, and physicists still don't know when the observable universe's expansion will stop.

Since the Big Bang, the universe was created and was a ball of heat because of its explosive reaction until it cooled down, but there is still remaining heat around the observable universe's halo that is still growing and expanding until infinity (∞) or so.

375,000 years after the Big Bang, An afterglow light pattern appeared, and thereafter, the galaxies and planets started to develop, the Milky Way Galaxy was developed and the birthplace where stars are born is in the nebulae. Nebulae are immense clouds of hydrogen gases where stars are born and develop. Why do stars develop on nebulae? Stars develop because of the hydrogen gas and rocks near the nebulae to form a birth star. Physics and the universe have a connection from the origins of the universe to the development of celestial objects like galaxies, planets, clusters, stars and violent explosions like supernovas and dangerous objects you can't get near of like a black hole.

Physics and its connection with the world we live in is invisible because people think physics is about equations and calculations, but it is actually more than that. Physics has connection with the time we have always in our devices, clocks and watches. And space, from the incredible sight of the most luminous Sun to the twinkling stars of the night sky. Physics is surrounded in anything we have in the world we live in, such as violent explosions.

As you see the image below, showing the Big Bang, and how it expanded in its timeline, how galaxies developed, our Milky Way, our solar system, stars, and planets as well.

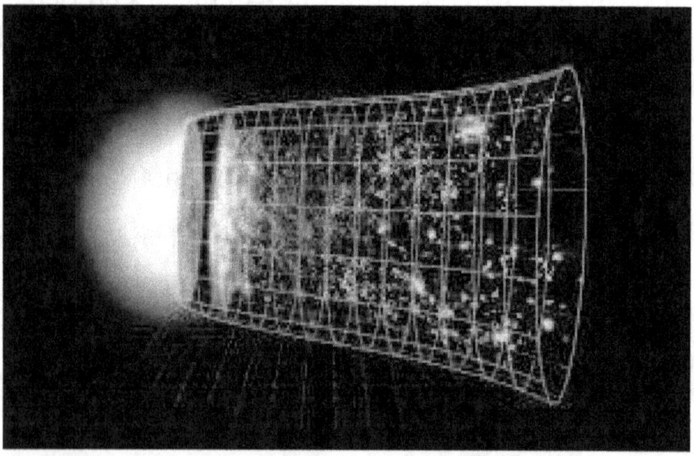

CHAPTER 2

EVERYTHING YOU NEED TO KNOW ABOUT PHYSICS & MATHEMATICS

Have you ever wondered how physics and mathematics are related to each other? Well, in this chapter, you will learn how mathematics is inside this astonishing field of study and how they are related. Physics' connection with mathematics is based on solving a problem that physicists are trying to find. Physicists use mathematical physics to calculate velocity, and other parts of physics' problems. For students at high schools, mathematical physics is a bit harder in everyday life, the reason behind this, mathematical physics has challenging problems that students can't even solve. Number systems in physics, have been easy and hard for primary schools and high schools but there is a reason why it's both easy and hard.

Real numbers are numbers that are not transcendental like Pi, or any other number. Example: [1; 2; 3… etc]

Non-real numbers are numbers that are transcendental

Example: [1,27365...; 1,56273164… etc]

Rational numbers are numbers that are fractional

Example: [½; ¼; ¾… etc] in percentages, it is 50%, 25% and 75%

Irrational numbers are numbers that are not fractional like square roots: Example: [√16] which is equivalent to four

Integer numbers are numbers that both positive and negative

Integers that start from +∞ to -∞ (positive infinity to negative infinity) like on a thermometer showing from above temperatures to below temperatures. There's also mathematical physics in extreme weather, if the Earth has below -180°C, all humankind and every life would die and time would freeze, this very dangerous and astonishing weather phenomena that has never happened before, it is also known as Absolute Zero, it has a decreased temperature of -273,15°C.

There is also mathematics in nature. Natural plants and its leaves have very beautiful mathematical shapes that are geometrically fascinating. The mathematician Fibonacci discovered that the sequence he created is a series of numbers that varies of those two preceding ones, it starts from 0+1 = 1, then 1+1=2, then 2+1 = 3, and so on.

There is some natural phenomenon known as the Fibonacci spiral, it is related to the golden ratio known as Phi (ϕ) in decimal version, it is 1.618… Mathematicians say that the Fibonacci spiral might be seen in art, nature and even finance. Why are they both related?

The Fibonacci spiral results in nature and astronomy. Like spiral galaxies, and pinecones which show a golden ratio or phi (ϕ) geometrical shape.

1:1.618 of the geometrical sunflower showing the Fibonacci spiral's astonishing geometrical existence of how mathematics has shaped nature, not only has shaped nature but the entire physics world. Even the greatest mathematical-natural object of them all, the Nautilus shell also has the same golden ratio spiral structure. Physics' relation with maths is really fascinating.

CHAPTER 2.1

PHYSICS & DIMENSIONS

EUCLID'S GEOMETRY AND THE PHYSICS BEHIND IT

Have you ever wondered how a dimension and geometry are related to time and space? What is a dimension? In mathematics and physics, a dimension is a minimum number of coordinates that are shown in a given amount of space. `

For example: A unidimensional (1D) shape is a line, just a horizontal line with one coordinate from point A to point B. Here is the image below; showing that the unidimensional shape has one vector or axis.

In ancient times, Greek mathematician Euclid discovered that shapes come from different dimensions and have a different number of coordinates, and there is an unknown dimension, somewhere beyond our universe, but that dimension cannot defy space but can defy time, it is also known as the fourth dimension.

People wonder what Euclidean space and geometry is, Well, what is Euclidean space? The term "Euclidean" is named after best Greek mathematician Euclid who discovered the geometry of dimensions. Euclidean space is the geometrical space that focuses on the topics of coordinates and vectors that are occurred on dimensions like 1D has the x-coordinate (horizontal axis).

One-dimensional space has one coordinate, and it has only two points and has an x-coordinate (horizontal), because its a line. Two-dimensional spaee has two coordinates, it has a length and width only, and it has a x and y-coordinate (both vertical and horizontal on both sides of the shape), proven in Euclid's geometry.

And finally, the most incredible dimension in the whole universe but it occurs in Euclidean space, 3D, shown in high-octane action movies, animation films and in real life. Three-dimensional space has three coordinates (length and width, which adds height as the third dimension). People think dimensions end on 3D, but there are one last dimension that show time, but it merges with the three dimensions of space (length, width and height), beyond Earth and the Milky Way. 3D has (x, y and z coordinates) which shows the three coordinates of space and matter.

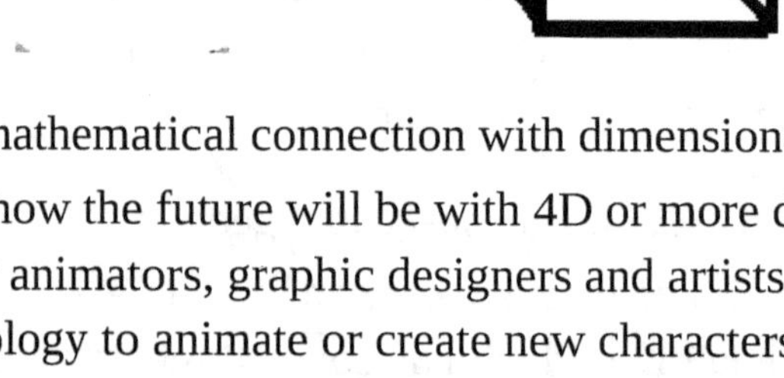

Physics' mathematical connection with dimensional geometry has "shaped" how the future will be with 4D or more dimensions. Computer animators, graphic designers and artists are developing 4D technology to animate or create new characters for their next films and video games. 2D was the first era of animation and gaming, and now there's a new era of 4D tech to shape our future.

Is there a fourth dimension or is it a myth? Well, there is no scientific fact about the Fourth Dimension, but it has been in the special theory of relativity that Einstein has been proving about, the spacetime curvature. The combination of space and time leads to the fourth dimension, three dimensions of space (length, width and height) and one dimension of time (which is a unknown coordinate, actually but it is considered as a "w-coordinate" from scientific research).

There is a four-dimensional shape that resembles as a cube but there is a cube inside of a cube, it is called a tesseract. A tesseract is a four-dimensional shape that resembles a cube inside a cube in its four-dimensional structure.

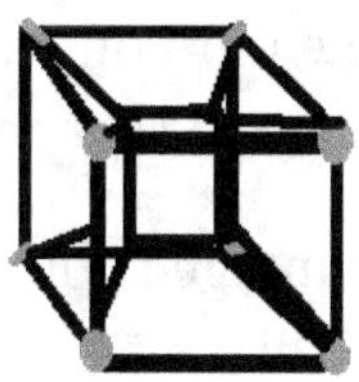

People believe that the Fourth Dimension was shown in the Marvel Cinematic Universe, and they never know that it theoretically exists. Well, yes, the fourth dimension exists because of geometry around us like a rectangular prism-like shack, or a cube-like stove oven. Geometry is the mathematics of shapes around us and helps mathematicians to solve angular (angle)

problems. Do you know that there is a secret that dimensions have when they extend from 3D to 4D and so on?

So, you ask yourself, what is the secret behind the dimensional extension or growth? The secret starts from 0D (zeroth dimension). 0D starts as a point and it has no coordinate, as the secret of the dimensional extension starts as the point duplicates to become two parallel points and they create a line and extends to 1D with x-coordinate which is a horizontal line.

A 1D shape or a line becomes two parallel lines as they create a 2D figure or shape, for example: a square, which has two parallel lines joined together to make a four-sided 2D shape, or in mathematics; we call it a quadrilateral.

Thereafter, a 2D shape or a square duplicates to two squares with joining lines that create the third dimension (also known as height) to originate a cube (a 3D shape).

Finally, the 3D shape has merged with the dimension of time (also known as the unknown 4th dimension). But in vectors, it has a w-coordinate like 3D has x, y and z coordinates showing height, width and length. Actually, the 4th dimension only shows nothing but time, which is a very eerie structure for the fourth dimension of time merging with the three dimensions of space.

The eerie secret of 4D showing a cube inside of a cube known as a tesseract, the 3D shape or a cube duplicates into two parallel cubes with joining lines creating the fourth dimension of time which leads to a tesseract, The dimensional extensions shows some kind of repeating pattern. How will 5D look like? In my opinion, 5D will originate from a four-dimensional shape like a tesseract duplicates into two tesseracts which will make a five-dimensional shape, which will look like a cube inside of a cube inside of a cube and finally, inside of a cube.

Here is the image below of the pattern:

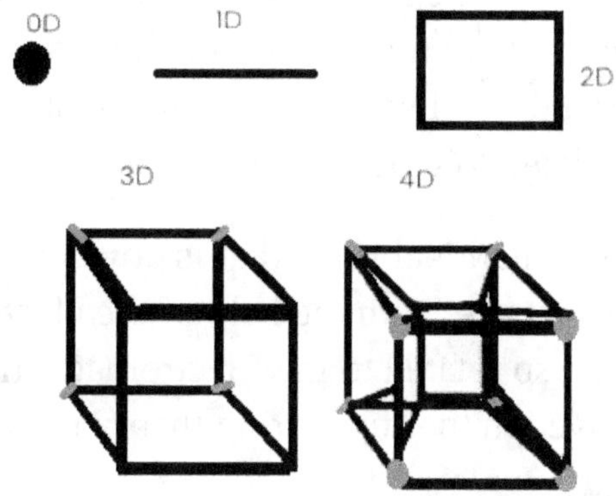

0D starts from a point, 1D, a line representing length, 2D a flat shape representing both length and width, 3D, a realistic shape adding height as the third dimension.

4D, adding time as the fourth dimension which is shown on Einstein's special theory of relativity about the spacetime curvature, proving to show that the third dimensions of mathematical space, was merged with single dimension of time. Dimensional coordinates are also shown in the mathematical physics of vectors. For example, the x-axis is horizontal, and the y-axis is vertical creating the dimensional figures of shapes.

This mathematical discovery was first proven by Ancient Greek mathematician Euclid, who discovered different types of dimensional spaces, like two-dimensional space which has length and width to create the 2-dimensional Euclidean figure.

Physics has also extra dimensional secrets to unravel in this incredible mathematical world of geometry. But this topic won't connect to Euclidean geometry. But it will connect with astrophysics and Beyond the Standard Model. There is a theory that Professor Edward Witten had been working on: the M-theory, is a scientific concept that proves that there are other universes that have fundamental 11-dimensional figures that is connected to a powerful force related to gravity, but super.

Supergravity is a gravitational force that is powered by electromagnetism in space, with the help of the electromagnetic radiation, which is also related to the Electromagnetic spectrum, it is said that is involved in the superstring theory to prove whether the 11-dimensional figures are true.

The superstring theory has an 11-dimensional formula: (seven-dimensional hyperspace + four common dimensions we have in both space and time (length, width, height and time). Hyperspace meaning that the universe is full of dimensional matter like a cube, spherical stars or even tesseract-like planets out there even scientists and physicists didn't observe in their space satellites.

There is an astonishing figure that has six dimensions on its spatial structure; it is the Calabi-Yau manifold that I saw on the string theory and M-theory. A six-dimensional figure is an astonishing shape ever observed in astrophysics, art, mathematics and the world. Why? Well, in string theory, a six-dimensional figure is shown to be "curled up" and showing weird curly patterns, meaning they are unobservable to the human eye, they are too microscopic.

A six-dimensional shape can add with the four common dimensions (three spatial dimensions and one time-based dimension) which are equivalent to ten-dimensional figure.

There is also a dimensional formula for velocity that resembles speed (distance divided by time to the power of negative one). $[D]\div[T]^{-1}$

For example; a car travels between 180 and 200 km/h (kilometres per hour) (rate of change) shown at a limited amount of velocity or speed.or between 100 and 180 m/s (metres per second) shown in of acceleration (amount of speed in a limited amount of time).

Dimensions has shown us the "points" of view in mathematical physics. Well, in the world of physics, geometry and dimensional insanity, we have learnt how Physics and Dimensions can really "shape" time and space both together and beyond the string.

CHAPTER 2.2

PHYSICS & DATA

DOES THIS INFORMATION "BYTE" YOU?

Have you ever wondered how many different types of data exists there? For example, mobile data you are using online, mathematical data that helps you find important key facts about the world of mathematics, data you analyze from research, or even computational data or computing data.

Data and physics have been each other since the dawn of computer era, not the data you use for your mobile, or for analysis. Find out how data and physics go hand in hand, one “byte” at a time. The data that was first made since the beginning of the Internet and the first Windows 1.0 computer back in 1985, 40 years ago. developed by Microsoft billionaire Bill Gates. Also, the first Apple Macintosh computer built in 1984, by Steve Jobs and other Apple developers. Here is the image below of the first-ever Windows and Mac computer:

Let's dive in to the physics of computational data, based from the SI units, which as the Systemé Internationalé d'Unités in French translated as "International System of Units" in English. The unit of data starts from bits. Bit is the building block of representing every computing data information and it can be 0 or 1 showing a binary system in the unit of data. Do you know how many bits create a byte?

Well, in the world of computing physics and data information of how computers work in the world of bits and bytes, Pixels are the building blocks of pictures that show visual information to people who perceive the image, Pixel is the combination of the words "picture" and "element", Picture shortened as "pix" and element shortened as "el", so it means the pixel is a picture element made of three colors that create a single image pixel, red, green and blue, creating luminosity and brightness on an image. There are even more bigger "bytes" you can handle in the world of 0s and 1s. A bit is also a binary digit made up of two possible numbers that create information through the computer's data systems, 0 and 1. Those binary numbers are not just ordinary numbers. Well, actually, this is how the first visual images looked like in the past, how rgb pixels look like in microscopic form, below.

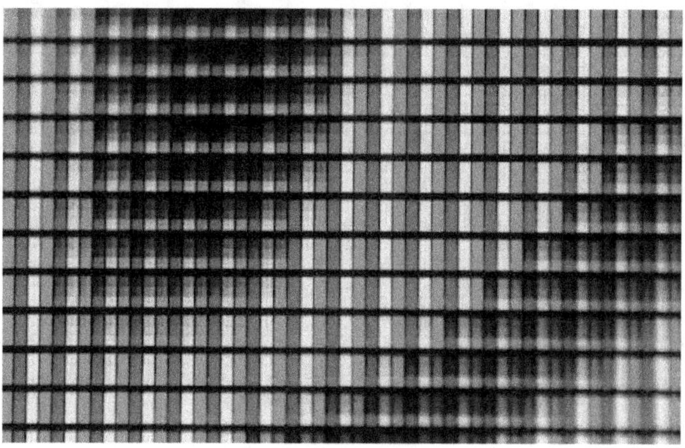

But binary digits that helped crack the German code and won the World War with the help of English mathematician and programmer Alan Turing. By meaning they are more bigger "bytes" to handle, I mean they are more units of data to learn and explore in the world of computing and its connection with physics.

In physics and data, there is a computational idea that helps you to program values or digits with the help of logic. It Is programming logic is a way of helping programmers to choose operators like + to add, it is like you are controlling your own computer. But sometimes, computers don't understand what unknown program other humans do and cause an error, which is called a logical error in computing.

A kilobyte enters through the system's hard drive, as we explore the physics behind kilobytes. A kilobyte is a unit of information or data that has 2^10 or a thousand and twenty-four bytes just to make a 100 by 100 pixelated image. The physics behind bytes starts as they multiply 10 bytes (equivalent to 80 bits) x 10 bytes x 10 bytes + 24 bytes to create a single kilobyte and it causes color depth growth of images.

Today, on modern video games or television screens, we have 236 gigabytes of internal storage to play games or based on over 1 million megapixels to create such a high-quality image. The actual physics behind the development from a bit to a byte to a kilobyte, etc, it is all the mathematical index or exponent of two.

A bit has just 0s and 1s, A byte has 8-bits, and a kilobyte has 1,024 bytes, why? The base, 2 to the power of 10 is equivalent to 1,024 bytes, even a megabyte (2^20) has 1,024 kilobytes (equivalent to 1,048,576 bytes). A gigabyte (2^30) has 1,024 megabytes (equivalent to 1,073,741,824 bytes).

In the metric system of data, there's a lot of secrets to unravel more bytes and its growth using the mathematical power of two. Well, on modern gaming consoles like PS5, they now use SSDs (solid-state drives) as their main disks for game storage which have 512GB and more to 1TB. Did you know, an unknown byte that is bigger than a gigabyte? Actually, everybody know the future of data. Well, not all people. Other people think that all storage drives end on the gigabyte era, but there's more to that.

A terabyte or TB (2^40) is a new, modern unit of data, for now, that has 1,024 gigabytes (equivalent to more than 109 trillion bytes), a mostly used unit of data for newly modern devices that have recent generations. Did this data "byte" you enough?

The image you see above, is our computer's heart or memory storage called the SSD (solid-state drive) it is used to store digital information on 512-gigabyte to 1-terabyte computers, and can also read and write memory in its motherboard, GPU (graphics processing unit) measured in kilobits per second, GPU is also used to measure the graphics performance of the computer like playing a highly-graphical game which has 60 FPS (frames per second) and CPU (central processing unit), it is the brain of the computer and controls every performance that the computer does, when running programs or apps from the operating system.

There are three different disk drives in the world of data; HDD, SSD and FDD. HDD stands for Hard Disk Drive, and it is one of the mostly used disk drives on the planet, on PCs, flash drives and on Smart TVs, but it has between 2 and 64 GB RAM which stands for Random Access Memory.

Random Access Memory in computing is a system where all files are stored and read any data in order and in the computer's given memory, CPU processor and a motherboard, the hard disk memory might be 464 GB or half a terabyte or more, not just on computers but on phones, and on other disk drive-built electronic devices.

SSD which stands for Solid-State Drive, it is the most powerful disk drive than HDD, it can have 512 GB to 1 TB or more, and it has magnets inside its coil to store files electronically with the power of Oersted's creation; electromagnetism. The SSD drive works magnetically and electronically, simultaneously.

SSD can also read, write and store files and can also be found on gaming consoles like Xbox and PlayStation for gamers to play any big-filed games that have 230 GB or more, SSD is more hard than HDD.

FDD which stands for Floppy Disk Drive, is a type of magnetic disk drive that reads, stores and writes digital information and it is removable like a USB (Universal Serial Bus) flash drive and other disk drives. Floppy disks are removable but they can store and move digital data or information between computers and other disk drive-built electronic devices. Floppy disks are now completely obsolete and when you insert the disk, it doesn't actually store, read and write digital data on modern computers like Dell OptiPlex.

Computers have two different types of digital data systems: decimal and binary systems. Decimal system (base-10) uses the power of 10, for example, a kilobyte (KB) equals to 1000 bytes (10^3).

Binary system (base-2) uses the power of 2, for example, a kibibyte (KiB) equals to 1024 bytes (2^10). “Bi” for binary, representing computer storage with the combination of 0s and 1s, while decimal system units like gigabytes which are most commonly used on computers which represents large amounts of data.

The bytes are growing faster on the IDC. The IDC is the International Data Corporation, it is a global organization where they operate all data systems on servers around the Internet and its connections globally. What I mean by growing faster? Well, data has been developing stronger, faster and better, from bits, to quettabytes.

A quettabyte (QB) is part of the International System of Units (SI), has 1024 ronnabytes or 1 quintillion gigabytes. The difference between decimal and binary units are SI prefixes:

In decimal units, the SI prefix examples like, kilo- which means 1000, or mega- has 1,048,576 bytes, and giga- which has 1,000 megabytes. In short, kilo, mega and giga (K, M and G).

In binary units, the SI prefix examples like kibi- which means 1,024 bytes, and mibi- has 1,024 kibibytes, and gibi- which has 1,024 mebibytes.

These are units that have same quantities, but has different exponential quantities, like I explained earlier, the decimal unit has the power of 10 and the binary unit has the power of 2 (consists of two possible numbers that are binary, 0s and 1s that creates a bit or a binary digit).

There is a computer system called R/W (read/write) on magnetic disk drives. How does a disk drive store data? Well, data can be stored through the R/W system. Data can be read by accessing digital information from the disk drive's magnetic storage system.

Data can also be written which is the process of saving and recording digital information on the disk drive's magnetic storage system.

Here is the image that shows how data transfers digitally between computers and other cloud systems with the help of physics:

Data is everywhere, it can also be transmitted through Ethernet cable with a high speed rate of 10 Megabits per second (Mb/s). In bit calculation, it has 10 million bits per second.

Yes, yes, we know, that data is involved in Wi-Fi (Wireless Fidelity) technology. People wonder what is Wi-Fi? Wireless Fidelity is a different local area network or LAN that has a frequency rate between 2.4 and 3 Gigahertz or GHz for short. Wi-Fi is part of the IEEE 802.11 (Institute of Electrical and Electronics Engineers) system.

Data is all around us, on our smartphones, laptops, TVs and Wi-Fi. But it's connection with physics has changed our world, from the very first pixelated games to the large amounts of bytes like gigabytes or yottabytes. Data and physics has not only shaped the digital world, but has shaped the future of IDC and the world.

CHAPTER 2.3

PHYSICS & RIVAL SYSTEMS

THE BATTLE BETWEEN METRIC AND IMPERIAL

Have you ever wondered about how engineers use units to measure the length of a wood plank in centimetres or in inches? Well, in the world of SI, there are two different unitary systems, The Metric System and the Imperial System.

The metric system was first developed by a group of French scientists in the late 1700s, that's why they call it Systeme Internationale d'Unites, today, while the Imperial System was developed by a group of British during the Industrial Revolution against America. That's why the US and UK use the Imperial System today, but the system has different imperial units between two countries, while the US uses the nautical mile, and UK uses the barleycorn unit.

What differentiates between these two unitary systems? The Metric System is the most commonly used unitary system from the International System of Units around the world except the United States of America. While the Imperial System, is being used in USA.

Why doesn[1]t the USA use the Metric System? Well, Since the American Industrial Revolution in the 1700s, were because of the financial problems that were caused by the British which led to the Imperial System of measurement., the unitary system that Americans use today. In America, people measure length in feet, yards or inches, they also measure capacity in gallons or fluid ounces.

People use the Imperial System to measure weight and mass in ounces, measure distance in miles, and measure distance divided by time which is equivalent to speed in a rate of miles per hour, only used in the United States of America.

The third president of the United States, Thomas Jefferson rejected the internationally-except-US Metric System because the currency system accommodated wealthy buyers and sellers at the expense of common farmers that had priorities inverted on, which really led to Imperial System from the British.

The physics behind these systems, are based on their different units and prefixes, for example in the Metric System, the prefix kilo- is equivalent to 1000 (1 multiplied by 1000), centi- is equivalent to 0.01 (1 divided by 100 or a hundredth of a unit), hecto- is equivalent to 100 (1 multiplied by 100), and finally, milli- is equivalent to 0,001 (1 divided by 1000).

The Metric System is a unitary system shown on the Systeme Internationale d'Unites (International System of Units) in English measures volume in litres, weight or mass in grams, length in centimeters, but physicists use it better than the US' Imperial System, because they use it to measure everything like speed (km/h), distance (km, m) or even how much time does it take for the rocket to land on the Moon, 384,000 kilometers away from Earth.

Physics and the metric system have been evolving because of the measurements people use to gather the data of how many units are the size of all objects, in the future, it might the measure the whole world, what will happen if there was no measurements?

Well, in ancient Egyptian times, Ancient Egyptians used their own body parts to measure the length of the pyramid, it sounds weird, right? For example: an eye is equivalent to 1 millimeter (a thousandth of a meter). There was a pharaoh named Horus, he had a symbolic eye, which is known today as the Eye of Horus,

What is the Eye of Horus? In Egyptian language, it is called an udjat. This symbol is not just an eye, but an hieroglyphical artifact that proves that it is a symbol of protection and defense, Hieroglyphs are symbols that represent linguistic communication to Ancient Egyptians, but for them, it's their historical language. The eye has mathematical rational measurements in its dimensions, for example: the udjat marking is a 1/64 (sixty-fourth from the eye of Horus` dimensions), the smallest part of the symbolic eye, shown below:

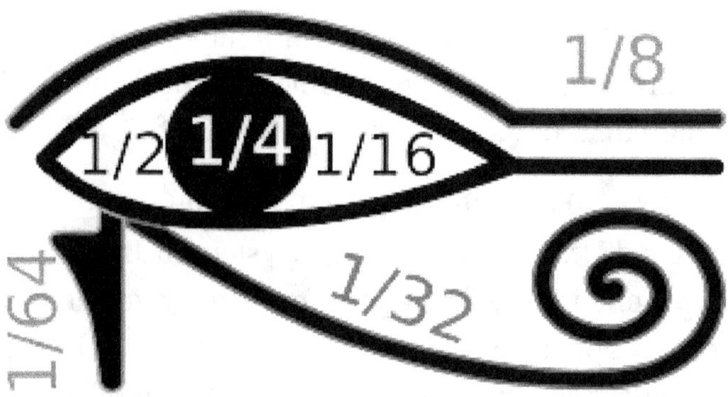

The mathematical physics behind this historical symbolic eye, is based on its fraction on each part of the eye, the eyebrow has 1/8 (eighth of the eye), the left eyelid has ½ (half of the eye), the right eyelid has 1/16 (sixteenth of the eye), the udjat marking under the eye on the left has a fraction of 1/64 (sixty-fourth of the eye), and the teardrop or an eye spiral under the eye on the right has a fraction of 1/32 (thirty-second of the eye).

The Imperial System is a unitary system that was first established in the UK. In America, people measure area in acres and hectares. They measure volume in gallons and so on. Physics' connection with the imperial system is based on the human body parts, like feet, or inch which is defined as the width of a man's thumb.

A unit that measures temperature in the metric system is Celsius, named after the astronomer and physicist Anders Celsius, who created the temperature metric unit. Even in the imperial system, a unit that also measures temperature, it's Fahrenheit named after the physicist Daniel Gabriel Fahrenheit.

Other people think °C and °F have same values and equivalences, But, actually these two different units have different values on two rivalry systems, the Metric and the Imperial System.

For example: 0°C (freezing point) is equivalent to 32°F, 100°C (boiling point) is equivalent to 212°F, and 37°C (normal body temperature) which is equivalent to 98°F.

Why these metric-imperial unit calculation different? Well, a Celsius unit multiplied by 9 divided by 5 and then add the answer by 32 equals to the Fahrenheit numeric unit. For example: 0°C x 9 which equals to 9°C divided by 5 add by 32 which is equal to approx. 32°C. Same thing happens between the distance units from the metric system and the imperial system, for example, like kilometers and miles (in abbreviated form: km and mi).

For example: 1 mile is equivalent to 2 kilometers from the metric-imperial unit conversion. Other physicists who commonly use the metric system convert the metric units to imperial units to gather information about how many miles per hour (mph) can a car be driven from Arkansas to Missouri?

There is another system in-between the metric and the imperial, it is known as the Avoirdupois System, it is a unitary system that is part of the Weights and Measures Bureau. Example units: ounces and pounds (.oz and lbs). There are 16 oz in 1 lb, according to the avoirdupois unit calculations. And there is a smaller imperial unit in the avoirdupois system, known as the grain (gr). There are approximately 7,000 grains in a pound.

In the world of mathematics and physics, there is more in the metric system. The metric system is also a decimal-based system of measurement, used to have measurements in the power of 10. There are metric prefixes like tera-, giga-, kilo-, mega-, milli-, micro-, nano- and pico- on other base units like bytes, meters, litres and grams.

In mathematical physics, we convert length from kilometers to meters (divided by 1000), moving from decimal point to the right or the left.

For example:

- 1 millimeter (1 mm) is equivalent to 0,001 m (divided by 1000).
- 1 kilogram (1 kg) is equivalent to 1000 g (multiplied by 1000)
- 1 microsecond (1 µs) is equivalent to 0,000001 s (divided by 1000000, a million)
- 1 megabyte (1 MB) is equivalent to more than 1000000 bytes (multiplied by 1000000)
- 1 picosecond (1 ps) is equivalent to 0,000000000000001 s (divided by trillion) a trillionth of a second

- 1 hectometer (1 hm) is equivalent to 100 m (multiplied by 100)

Here is the images below showing the SI prefixes and its decimal values or multipliers:

Table 5. SI prefixes

Factor	Name	Symbol	Factor	Name	Symbol
10^{24}	yotta	Y	10^{-1}	deci	d
10^{21}	zetta	Z	10^{-2}	centi	c
10^{18}	exa	E	10^{-3}	milli	m
10^{15}	peta	P	10^{-6}	micro	µ
10^{12}	tera	T	10^{-9}	nano	n
10^{9}	giga	G	10^{-12}	pico	p
10^{6}	mega	M	10^{-15}	femto	f
10^{3}	kilo	k	10^{-18}	atto	a
10^{2}	hecto	h	10^{-21}	zepto	z
10^{1}	deka	da	10^{-24}	yocto	y

The Imperial prefixes are not shown decimally, but shown rationally, in a fraction way, for example ½, ¼, 1/8, 1/16, 1/32 and 1/64.

A pint is a unit of measurement used in the imperial system to measure volume, 2 pints is equivalent to 1 ¾ liters. A gallon is equivalent to 3 liters. A fluid ounce (fl. oz) is a US imperial volume unit used for measuring liquids and its equivalent to 29 ½ milliliters (0,029 liters). A liter's equivalence with the fluid ounce units (a liter is equal to more than 32,000 fluid ounces).

In imperial units, meters are squared in area measurement. They are square meters, square inches, square feet and acres, that are have the same unitary standard like centimeters and millimeters.

There is more to physics and the rivalry systems like metric vs. imperial; let's focus on the non-metric but SI-accepted units that around us, like time and other non-metric units.

There are units like minutes, or hours. These units are base-60 meaning they are sexagesimal similar to a decimal that is base-10.

Everybody knows that a minute has 60 seconds, in the world of mathematics, time and physics, a minute and a second have different durations, but they do have the same standard.

For example; a second is divided into four rations of a single duration

- a quarter of a second is equivalent to 250 milliseconds (1000÷4 = 250)
- a third of a second is equivalent to 750 milliseconds (1000÷3 = 750)
- Half of a second or known as the semi-second is equivalent to 500 milliseconds (1000÷2 = 500)
- Obviously, a whole of a second is equivalent to 1000T milliseconds.
- Quarter of a minute is equivalent to 15 seconds (60÷4 = 15)
- Third of a minute is equivalent to 20 seconds (60÷3 = 20)

- Half of a minute or a semi-minute is equivalent to 30 seconds (60÷2 = 30)
- Obviously, the whole of a minute is equivalent to 60 seconds

Physics' connection with the metric system is also resulted in everyday life; for example; when people are about to prepare their own tea every day they simply measure two to five tablespoons (tbsp) of sugar when making tea.

There is also one and final metric unit of temperature, and it is called the Kelvin scale, that was first proposed by the British scientist Lord Baron Kelvin, that 1° K which is equivalent to 1° C.

The difference between degrees Celsius (°C) and degrees Kelvin (°K) is that the Kelvin scale starts from the coldest possible temperature that in theory can freeze time and space, known as absolute zero, which is equivalent to -273.15° C, and in Kelvin scale it is 0° K. But they have same values or increments; for example; 10° K equals as same as 10° C, like I explained earlier about 1° K is equivalent to 1° C.

As for the Celsius scale, starts from 0° C, as it is known as the freezing point of water. As for the Fahrenheit scale, the freezing point of water, starts from 32° F, the absolute zero scale, it is equivalent to approximately -460° F. And finally, the Kelvin scale's freezing point of water begins at a scale of 273° K, as for the absolute zero scale, obviously begins at 0° K.

The boiling point of water starts from 100° C, in Fahrenheit scale, it is equivalent to 212° F, and in the Kelvin scale, its boiling point begins at a scale of 373° K.

Here is the Celsius-Fahrenheit-Kelvin scale diagram showing data starting from absolute zero to the hottest temperature possible:

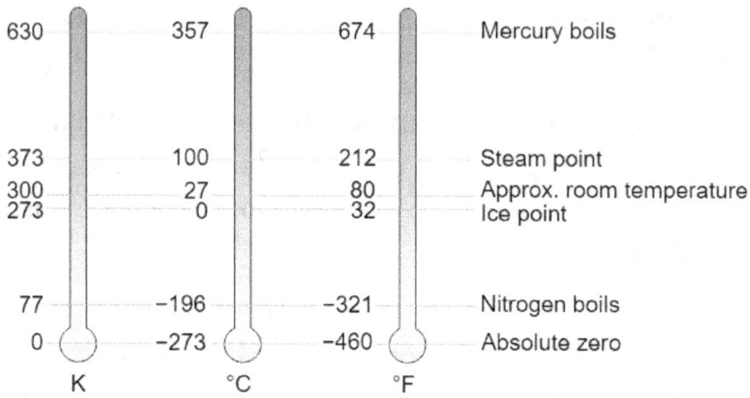

We are now done with the past and present of the metric system, we are now travelling to the future of how physics and the system of measurement that the world uses today, known as the metric system will help us solve problems or some other everyday things we people do. In the future of the SI, there will be moretechnological advancements in physics based on how a kilogram or a milligram develops into different constants, maybe a kilogram will be 1000.82 grams or more, a milligram can be 0.00128 grams or less.

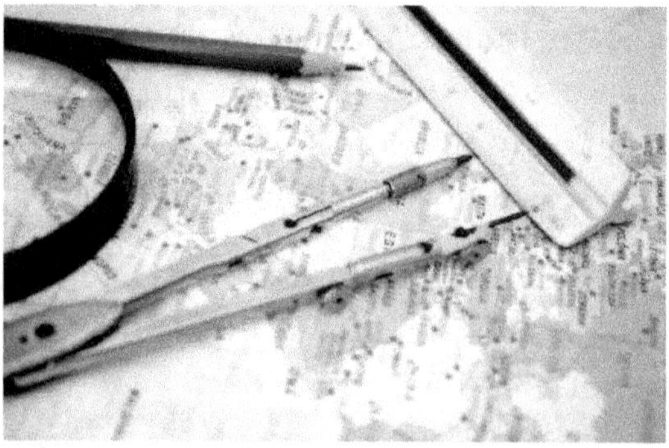

CHAPTER 2.4

PHYSICS & UNITS

HOW WE USE THESE WITH THE HELP OF PHYSICS

Have you ever wondered how many units we have in the world? Well, in this chapter, we are going beyond the metric system, Physics and units have shaped the mathematical world of metrics and imperials, since the development of the early metric system in France. There are units of measurement, units of speed, units of distance, units of volume, units of force, units of everything that our world has for us.

There are seven SI base units in the world of metrological and mathematical physics, physicists, metrologists and mathematicians use to gather data from sources or measure objects from space or anywhere.

Here are the International System of Units' base units:

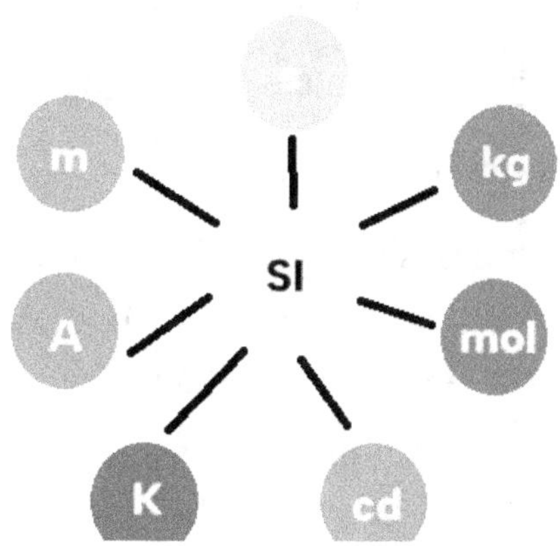

- meter (m) for length, equivalent to 1000 millimeters which leads to a meter
- second (s) for time, equivalent to 1/60th of a minute, 1/86400th of a day
- kilogram (kg) for mass, equivalent to 1000 grams
- kelvin (K) for temperature, equivalent to 1° C in centigrade
- mole (mol) for molecular mass
- candela (cd) for luminous intensity, measuring different types of brightness
- ampere (A) for electricity, similar to volts and watts

We people use measurement devices to measure how many centimeters or other units can the object be,

For example:

- Rulers, measuring tapes (centimeters, meters, millimeters)
- Thermometers (kelvin, fahrenheit, celsius)
- Rain gauge (millimeters of precipitation)
- Clock (second, minute, hour)
- Candela focuses on measuring luminous intensity, 1 lumen per steradian, a steradian is a unit of a solid angle depending on how many degrees

People wonder what are units? Units are standards used to compare different measurements of physical quantities. They are also derived units meaning they have rates of change, for example: Speed is equivalent to distance divided by time (kilometers per hour, km/h).

Derived units are actually units of measurement that are used to combining base units with multiplication and division. They are used to express physical quantities that are not fundamental in the International System of Units (SI), for example: area, volume, velocity and acceleration. They are also called "rates of change" units in mathematics.

Examples:

- Area: Measured in square meters (m^2), multiplying length by width ($l \times w$)
- Volume: Measured in cubic meters (m^3), multiplying length by width by height ($l \times w \times h$)
- Velocity/Speed: Measured in kilometers per hour (km/h), dividing distance (kilometers) by time (hour)
- Force: Measured in Newtons (N) or kilogram multiplied by meters per second squared ($kg \cdot m/s^2$) multiplying mass (kilogram) by acceleration (meters per second squared) also shown on Newton's 2nd Law of Motion proving $f = m \times a$
- Acceleration: Measured in meters per second squared (m/s^2), dividing distance (meters) by time squared ($t \times t$)

What are units without metric prefixes? Let's explore behind the metric unitary calculations of these derived units, for example; 1 hertz (hz), which is a unit of frequency is equivalent to 1 cycle per second (1/s)

- A hertz is equivalent to 1,000 millihertz; or 1000 × 0.001 hertz to be precise
- A kilohertz (KHz) is equivalent to 1,000 hertz or 1000 cycles per second
- A megahertz (MHz) is equivalent to 1,000,000 hertz or a million cycles per second
- A gigahertz (GHz) is equivalent to 1,000,000,000 hertz or a trillion cycles per second
- A terahertz (THz) is equivalent to 1,000,000,000,000 hertz or a quintillion cycles per second

Radio stations transfer communication signals around the world using Frequency Modulation or FM. They use units of frequency mostly megahertz (MHz), Hertz and bytes are similar but different. Hertz are units of frequency, radio broadcasters use to transfer signals, while bytes are units of digital information, the devices use to transfer digital data from a network on any source.

Candela are units of measurement, and it is one of the base units in the International System of Units or SI for short, based on luminous intensity, like I explained earlier, that a candela is equivalent to 1 lumen per steradian, a luminous flux divided by a solid angle degree.

A steradian is also equal to radian multiplied by a radian which makes radian squared (rad^2). Lumens are units of measurement that measures visible brightness emitted from a source, Lumens per angle when derived into a candela.

Candela is named after the item that we use every day to see things in the dark, a candle. A candela unit can be linked to the human eye by perceiving different intensities of light.

As I explained before, physics and units have shaped the mathematical world of the metric and imperial system.

Now, we are exploring the imperial base and derived units:

- For length, base units are foot, inches, yards and miles
- For volume, the base units are pints, gallons and fluid ounces.
- For mass, the base units are ounces and pounds

These base units are also involved in the imperial derived units like the metric derived units.

For example:

- density is equivalent to mass per volume, in the metric derived units, it is kilogram per cubic meters, then in the imperial derived units, it is ounce per cubic foot.
- Velocity is equivalent to distance per time, in the metric derived units, it is kilometers per hour, then in the imperial derived units, it is miles per hour.
- Force is equivalent to mass × acceleration, in the metric derived units, it is kilogram multiplied by meters per second squared.

From the early metric system to the sizes of our natural and futuristic world, physics and units have a connection to the International System of Units or the metric system and also the

imperial system or the British Weights and Measures System. Physics and units that are base or derived has also shown us mathematical quantities of these SI units, for example, a hertz is equivalent to 1 cycle per second.

CHAPTER 2.5

PHYSICS & SPEED

SOME LIGHTNING-FAST FACTS

Does the Flash even exist, and how can distance divided by time lead to speed? Speed is all around us, why? Physics is the reason why speed is everywhere, from how light is faster than anything to the speed timelines of the Earth's rotation around the Sun which takes one year to finish its revolution. Physics has a need for speed by using the unit of distance or length, a meter divide by a time unit known as a second. Do you have a need for speed? Fasten your seatbelts as this chapter travels you to the journey of how physics and speed work together!

What is speed? Speed is a derived unit made up of distance divided by time (meters per second or kilometers per hour). People who drive cars look at the odometer to check whether the car is speeding up from 170 to 200 km/h. Look at how physics has revolutionized the world of cars and runners with speed.

Let's get away from facts, and move on to the fictional stories of physics and speed together. Based on the DC characters, the Flash, is a fictitious character that represents the power of speed and time. The Flash aka Barry Allen faced a hard time as he was struck by lightning, and had gained super speed, and he used his powers to save his city from villains, he used his powers to travel in time and the distance was too fast. The Flash is an example of a physics-inspired DC Comics superhero.

People wonder what is faster than anything we have on Earth? Well, that answer is very simple, light. Light is faster than anything possible or fast in the world. But gravitational lenses from a black hole can bend light. From Einstein's energy-mass equivalence or $e = mc^2$, there is a "c" on the famous equation. That c is also known as the speed of light which is how much light faster at a high-speeding rate . Speed of light is equivalent to 300,000,000 or 299,792,458 meters per second in a vacuum.

But in fiction, the Flash is faster than the speed of light, meaning he runs at a super-speeding rate of 400,000,000 meters per second. Meaning time slows down, and there's a lot of infinite energy building up, and that's how he's able to travel back in time.

There is an impossible question that people are wondering so far, can speed really help you travel to the future? In Einstein's General Theory of Relativity, it is theorized that speed and gravity can help you or any flying object rotate around the Earth faster, which makes time go faster and maybe help you travel to the 2030s. This unproven theory is called "gravitational time dilation". It works from the invisible force of gravity above the spacetime curvature of the Earth.

In Physics and Speed, there is acceleration. People wonder what is acceleration?

Well, acceleration is a quantity that shows us how the speed of an object changes and acceleration is defined as a meter or distance as per unit of time or a second which is 1/60th (sixtieth) of a minute, and it is a metric or SI derived unit of speed. Acceleration was also explained by theologian and physicist Isaac Newton on his 2nd law of motion that force is equivalent to mass multiplied by acceleration, or $f = m \times a$ in an abbreviated form. In acceleration, when the car speeds up, the acceleration increases and moves forward and the velocity moves forward, but when the car slows down, the acceleration decreases and moves backward, while the velocity continues to move forward. When the car slows]down, it creates an opposite or inverse version of acceleration, which is known as deceleration.

- Acceleration speeds up
- Deceleration slows down

How does the speed equation work? And why is it involved in our everyday lives?

- Speed is equivalent to distance per unit of time
- Or it may change from speed to distance, Distance is equivalent to speed × time
- Or, it may change from distance to time, Time is equivalent to distance as per unit of speed

That is why in fiction, The Flash travels back in time with the help of super speed in a highly amount of distance. The Flash's super speed can also solve Einstein's energy-mass equivalence such as $e = mc^2$ from high distance, building up infinite electrical energy, and travels back or forward in time, and his mass as well.

The speed equation may involve in our daily lives, for example:

- The plane travels for 10 hours, with a distance rate of 160 km, what is the speed that the plane will reach to the New York International Airport?
- How we people calculate the speed equation from the problem unsolved above: Speed is equivalent to distance divided by time
- S = d/t; Distance = 160 km, Time = 10 hours, 160 divided by 10 is equivalent to 16 km/h. The speed is actually 16 km/h.

The works of the speed equation has been represented on the Flash, distance, time and even light faster than anything in our universe.

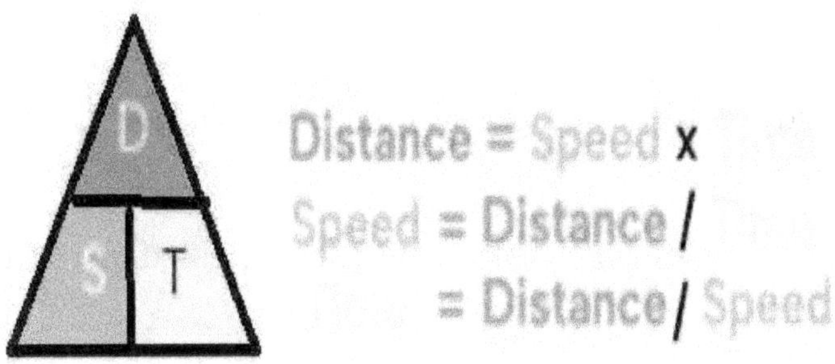

Physics and the lightning-fast world of speed, has a connection between distance, time, fiction and acceleration, and our everyday lives. From walking, running, cycling and sprint races on the Olympic Games, from the first sprint race win of Jesse Owens in 1936 to the future of fast racers today like Akani Simbine from South Africa, Usain Bolt from Jamaica, Faith Kipyegon from Kenya, and Joshua Cheptegei from Uganda. Speed has influenced their lives, and has inspired them to run and run until they achieve their biggest goals, even though Usain Bolt is retired.

He might be slower than the Flash, in fiction, even if the Flash existed. But in reality, Usain Bolt is the fastest human ever that a high speed rate of 12,4 meters per second, in 2008 at Berlin, he ran at a distance in at a time rate of 9,58 seconds.

- Speed = 12.4 meters per second (m/s), Distance = ?, Time = 9,58 seconds
- Distance = speed × time, d = 12,4 × 9,58 = 118,79 meters
- The distance is 118,79 meters

Usain Bolt broke the laws of physics and speed, and made history of being the fastest man, but, not faster than any other athlete on Earth. Physics and speed has shaped the world of time and distance, from milliseconds to kilometers.

The comparison between Bolt and the Flash, if he existed:

- Usain Bolt runs at a rate of 12.4 meters per second
- The Flash runs at a rate faster than the speed of light, which is more than 299,792,458 meters per second
- Usain Bolt is the fastest man but they are other runners who are faster than him in reality
- The Flash is the fastest DC Comics character ever created and inspired by speed physics in fiction
- These two are incredibly fast, both in reality and fiction, But the Flash has high speed that Usain Bolt in fiction.

What's next for speed and physics? There are some technological advancements in transportation, for example:

- Hyperloop technology, which was developed and originated in Asia, Europe and North America.

- Hypersonic aircraft, aircraft that levitates with super magnets with high magnetic force on air, which is still in development, that will develop some kind of super speed, maybe faster than the speed of light.

These newly technological advancements will be developed, based on the DC character, The Flash, or other speedy superheroes to develop new speed and saving energy more efficiently on those new devices.

Physics and speed are also showing us a brighter and more technological future that deals with new technology on fasttransportation like maglev (magnetic levitation) trains powered by magnets.

These are examples of how hyperloop transport and hypersonic aircraft will look like in the future:

Hyperloop technology is a different type of transportation technology that builds a vision of high-speed vehicles where they travel on clear tubes that help them accelerate faster with the help of magnetic levitation where the rails are filled with neodymium magnets, the most powerful magnets on Earth, maybe not powerful than the Earth's huge magnet inside its surface.

What about hypersonic speed? Hypersonic speed is a defined as a distance as per unit of time, faster than the speed of sound, neededon planes with hypersonic technology. A vehicle that has hypersonic speed is the Mach 5, has flies at a rate of 3.80 m/s, while the normal speed of sound in dry air is about 3.43 m/s. There is another different type of speed, digitally, and that is Internet speed.

Internet speed is a maximum capacity at how much data is being transferred slowly or faster. Bandwidth is the maximum capacity that shows how much data is being transmitted on internet-connected devices like computers, or smart TVs, at a data rate of gigabit per second, or megabit per second, (Gbps or Mbps).

On the image on the previous page, it shows someone holding the phone that shows how much the data is being transferred, representing network speed, and on the right of the rate, it shows how much data is being transmitted or received, representing network bandwidth.

In history, the Italian physicist Galileo Galilei was the first to measure speed and observed about the speed equation:

- v = d/t (Velocity/Speed = Distance ÷ Time

We learnt that speed and physics have shaped our universe, not just our universe, but also our everyday lives, from great athlete Jesse Owens' historic 1936 victory to the future of speed transportation with the power of magnetic force. We learnt that in fiction, the Flash is faster than the speed of light, but in reality, matter can't reach faster than the speed of light, if it had an infinite amount of energy like the Flash. Speed and physics has struck lightning through our curiosities and gained more knowledge about speed's connection with physics.

CHAPTER 2.6

PHYSICS & PIES

THE USE OF THE 3.14 NUMBER AND EINSTEIN'S "FAVORITE" TRANSCENDENTAL NUMBER

Have you ever wondered why is Pi celebrated on Albert Einstein's birthday? Physics and the famous mathematical constant Pi have a connection that you can have a piece of Pi baked by your grandmother every week. Pi is equivalent to 3.14159……

Pi (π) is the ratio to the circle's circumference to the diameter. A circumference is the outer layer of the circle around its shape. A diameter is a horizontal line inside a circle divided to into two radii. Pi can be found on the circular pies, that you eat every day or on holidays. Pi can also be found on many circular objects.

Pi is a transcendental number. What is a transcendental number? A transcendental number is a mathematical number system that is not algebraic, and has no roots like square roots or cubic roots.

Why is Pi celebrated on Albert Einstein's birthday? Pi is celebrated because of its number 3.14 which represents the date March 14th, it is Pi Day. Albert Einstein's birthday and the death anniversary of the late professor Stephen Hawking are also on March 14th, making it Einstein's favorite transcendental number. Pi Day was made in the USA and the holiday was made in 1988 by Professor Larry Shaw.

Let's head back in time from the origins of Pi, this famous equation today, was first calculated by Greek mathematician Archimedes, who circumscribed polygons into circles to the value of pi which he called it in a Greek symbol "π". Pi in fraction is equivalent to 22 divided by 7.

Physics and Pi have a connection between circles and the world around us. Pi's connection with physics is the mathematics behind it, from the first ever calculation of the great Greek mathematician Archimedes, representing the number 3.14 to the celebrations of this special constant, Albert Einstein's birthday and late professor Stephen Hawking's death anniversary. Do you want to eat a piece of "Pi"? If you know, what I mean.

CHAPTER 3

EVERYTHING YOU NEED TO KNOW ABOUT PHYSICS & ATOMS

Have you ever wondered what matter is made of? Well, the universe is made of matter, but matter is made up of atoms. Atoms are the fundamental but tiny building blocks of matter.

Atoms are made up of small particles, electrons, neutrons and protons.

- Electrons are particles that have negative charge and repel around protons and attracts with other electrons, as they say opposites attract, but similarities repel.
- Neutrons are particles that have neutral charge.

  While protons are particles that have positive charge and repels around electrons and attracts with other protons.

While the middle part of the atom, is called the nucleus. The nucleus is the control system of the atom, and attracts both particles in its surface that leads it into an atom.

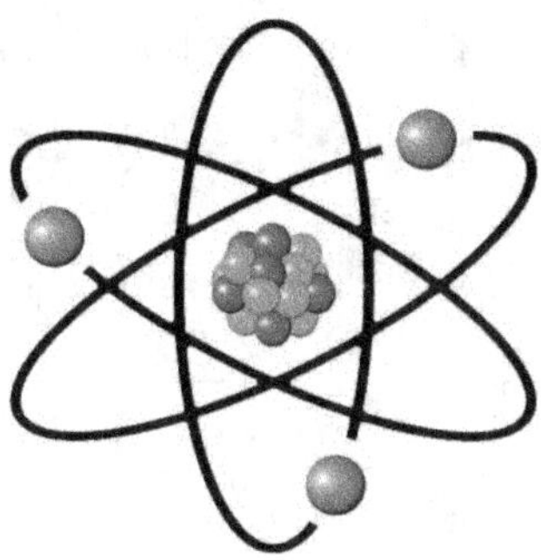

The atom is not just a matter that creates particles, but its the fundamental building block of our universe, and everything around our world.

But theoretical physicists believe that they are atoms out there that are "enemies" with actual matter, these deadly atoms are called antiatoms, which creates antimaterials, antimatter and anti-everything in other planets. Antimatter opposes matter. If these two enemies collide, it will cause a big cataclysmic explosion.

Atoms are also involved in Russian physicist Dmitri Mendeleyev's creation, which is called today, the Periodic Table of Elements, the table has different groups (horizontal part) and periods (vertical part).

The Periodic Table of Elements has 118 different elements that have symbols, names (other names are named after countries and famous physicists) and different atomic weights from the abundant element Hydrogen to the last element Ununoctium.

The Periodic Table of the Elements

Group→ / ↓Period	1	2	3	4	5	6	7	8	9	10	11	12	13	14	15	16	17	18
1	1 H																	2 He
2	3 Li	4 Be											5 B	6 C	7 N	8 O	9 F	10 Ne
3	11 Na	12 Mg											13 Al	14 Si	15 P	16 S	17 Cl	18 Ar
4	19 K	20 Ca	21 Sc	22 Ti	23 V	24 Cr	25 Mn	26 Fe	27 Co	28 Ni	29 Cu	30 Zn	31 Ga	32 Ge	33 As	34 Se	35 Br	36 Kr
5	37 Rb	38 Sr	39 Y	40 Zr	41 Nb	42 Mo	43 Tc	44 Ru	45 Rh	46 Pd	47 Ag	48 Cd	49 In	50 Sn	51 Sb	52 Te	53 I	54 Xe
6	55 Cs	56 Ba		72 Hf	73 Ta	74 W	75 Re	76 Os	77 Ir	78 Pt	79 Au	80 Hg	81 Tl	82 Pb	83 Bi	84 Po	85 At	86 Rn
7	87 Fr	88 Ra		104 Rf	105 Db	106 Sg	107 Bh	108 Hs	109 Mt	110 Ds	111 Rg	112 Cn	113 Nh	114 Fl	115 Mc	116 Lv	117 Ts	118 Og

Lanthanides	57 La	58 Ce	59 Pr	60 Nd	61 Pm	62 Sm	63 Eu	64 Gd	65 Tb	66 Dy	67 Ho	68 Er	69 Tm	70 Yb	71 Lu
Actinides	89 Ac	90 Th	91 Pa	92 U	93 Np	94 Pu	95 Am	96 Cm	97 Bk	98 Cf	99 Es	100 Fm	101 Md	102 No	103 Lr

CHAPTER 3.1

PHYSICS & PERIODIC TABLE

A LOT ABOUT ELEMENTS, NOT FIRE, WATER, AIR AND EARTH

Have you ever wondered how many metals are involved in the Periodic Table of Elements? Yes, I explained the Periodic Table in the previous chapter, but, this time, I am going to explain more about it and its elements and the inner workings of this Periodic Table of Elements.

Like I explained earlier, the Periodic Table of Elements has 118 different elements from the first element Hydrogen to the 118th element Ununoctium. This incredibly amazing table was originated by Russian physicist Dmitri Mendeleyev in the late 1800s.

You may be wondering, what do I mean by there other elements named after physicists and countries? For example:

- (Am) Americium named after the country America, or...
- (Sg) Seaborgium named after the physicist Glenn Seaborg
- (Mv) Mendelevium named after the creator Dmitri Mendeleev or Mendeleyev.
- (In) Indium named after the country India.

Weird, right? Now you see what I mean. These other elements are named after countries and physicists. Elements play an important role in the Periodic Table of Elements. Not elements that our connected to the astrological Zodiac which Fire, Water, Air and

Earth, not those kind of elements. But chemical elements. Chemical elements that have different inner workings, for example: The 26th chemical element of the Periodic Table, known as Fe or Iron, has the inner working of transporting blood around our bodies through our veins. Iron and other chemical elements plays an important role in our life too.

For example:

- Oxygen (O, atomic number 8), plays an important role in our daily life, too, in botany, during respiration, plants exchange carbon dioxide (from human exhale) to oxygen (human inhale), and it is also involved as the 21% of our earth's atmosphere with 78% of nitrogen in our earth's atmosphere to breathe
- Hydrogen (H, atomic number 1), also plays an important role in our daily life, it might be a flammable gas, but it is also helpful as a diatomic molecule when combined with one oxygen molecule to become the most healthiest drinks ever made by nature, H^2O, or Water, by hydrating our bodies whenever we need it.
- Calcium (Ca, atomic number 20), also plays an important role in our daily life, it helps us to be energized from consuming medicine or eating greens that have calcium. Also, calcium can also be found in our bones.
- Magnesium (Mg, atomic number 12), also plays in important role in our daily life, by battling deficiencies in our bodies.
- Zinc (Zn, atomic number 30), also plays an important role in our daily life, to battle against illnesses in our bodies, and also a chemical used in electricity.

- Copper (Cu, atomic number 29), doesn't play an important role in our daily life, but plays an important role n electricity. It is used to conduct electricity within insulated wires because of its high resistance measured in ohms (Ω).

But other chemical elements can be poisonous and radioactive compared to other elements like Arsenic (As, atomic number 33). Also, Silicon (Si, atomic number 14) can be used to semi-conduct electricity based on the solar panels. The Periodic Table has incredible inner workings and fascinating chemical elements.

The Periodic Table has three different variants of elements:

- Semi-metals (Silicon, Arsenic and Aluminium)
- Non-metals (Hydrogen, Neon and Oxygen)
- Metals (Zinc, Iron and Copper)

These elements can merge with other elements, it is called a chemical compound. Chemical compounds are combinations of two or more elements like Fe^2O^3 (Iron Oxide, has two iron molecules and three oxygen molecules) More oxygen, less iron.

There is a history of the Periodic Table, it all started from the Big Bang (13,8 billion years ago), were the world was nothing but darkness but filled with hydrogen gas which makes it the first element in the Periodic Table.

Helium (He, atomic number 2), might be a flammable gas, but it can be used to inflate balloons, or change our voices from normal voice to high-pitched voice.

Let's explore the elements inside our Periodic Table and how they work in our daily life, starting with:

- Hydrogen (H, atomic number 1), the first element in the Periodic Table of Elements, flammable gas, abundant element in our universe, and it is the creation with Water as a chemical compound with Oxygen (H^2O)
- Helium (He, atomic number 2), flammable gas, used in our daily life, for cooking when there is no electricity, or used for inflating party balloons, and changes the pitch of your voice, when inhaling it.
- Lithium (Li, atomic number 3), flammable gas and element, found inside lithium-powered batteries, and was first created by French physicist Antoine Lavoisier.
- Carbon (C, atomic number 6), most used element in daily life, used to manufacture graphite pencils, and also found in our bodies, when we are exhaling carbon dioxide (dry air) (CO^2) as a chemical compound with combined with two oxygen molecules.
- Nitrogen (N, atomic number 7), flammable gas, it is also involved in the atmosphere at a percentage of 78% in the air. Nitrogen can cool down below -170° C which creates liquid nitrogen.
- Oxygen (O, atomic number 8), also flammable gas, and it is also involved in the atmosphere at a percentage of 21% in the air.
- Sodium (Na, atomic number 11), it is a chemical element when combined with poisonous gas known as Chlorine to become salt (NaCl). But salt in real life, is less poisonous, it has more sodium, less chlorine based on its low pH.

Look at the image below, when salt becomes a crystal known as halite or sodium chloride in scientific terms.

The Periodic Table has different metals and gasses:

- Alkali metals
- Alkali earth metals
- Noble gas
- Halogens

They are other elements named after physicists and countries

For example again:

- Einsteinium (Es, atomic number 99), named after German physicist Albert Einstein

The most poisonous and deadliest elements I know, it is Mercury or Hg, atomic number 80. Mercury is a very toxic substance, it is also found inside thermometers to measure the temperature of a person or of an object. Mercury's chemical symbol "Hg" comes from the word "hydrarygrum".

Physics and the Periodic Table can lead to unexpected events, from the first ever Atomic Bomb to the deadliest Gas Tanker Explosion in Boksburg in late 2022. These events are happening because of deadly atoms that deform our structures in our everyday life.

Without Dmitri Mendeleev, the Periodic Table wouldn't been existed and we, people would never know the 118 different elements, those elements that are very essential in our daily life and elements that are destroying our Earth, but they are made from forces of nature. Even, elements like Neon (Ne, atomic number 10), is a noble gas, and it can make a high quality of luminous intensity, meaning for example; people use Neon lights for birthday parties, or for shops as signs like Welcome.

The incredible Periodic Table of Elements has 118 elements, divided into four components: alkali metals, alkali earth metals, halogens and noble gasses.

Let's discover the elements that in the alkali metals:

- Lithium (Li, atomic number 3), inside lithium-powered ion batteries
- Sodium (Na, atomic number 11), inside salt when combined with the poisonous gas Chlorine (Cl, 17)
- Potassium (K, atomic number 19), found inside our bananas and found inside sugary foods when merged with acesulfame, to become acesulfame-K or acesulfame potassium.

- Rubidium (Rb, atomic number 37)
- Caesium (Cs, atomic number 55), found inside atomic clocks to measure the rotation of our Earth's acceleration and speed.
- Francium (Fr, atomic number 87), named after the country France.

Noble gasses like Helium, Neon and other gasses are very flammable and can be involved when changing the pitch of our voice.

- Helium (He, atomic number 2), a very odorless and flammable gas that can be used to inflate balloons and change the pitch of our voices.
- Neon (Ne, atomic number 10), a very odorless and flammable gas but it is used on neon lights that are signs you see on shops.

Magnesium and calcium (Mg & Ca, atomic numbers 12 & 20) are examples of alkaline earth metals that can also help boost your immune system and help increase your brain abilities from products you see on TV like Centrum.

Have you ever wondered what halogens are? Halogens are a group of chemical elements that have nonmetal structures or gassy structures like Fluorine and Chlorine (F and Cl, atomic numbers 9 and 17). The element or atomic number 117 in the Periodic Table, also known as Ununseptium is also a halogen, but scientists and data researchers haven't discovered it that much yet.

Well, chemistry is all around us. From developing atoms, to the daily explosions that happen every day, every week, every month and especially, every year to be precise.

For example:

The supernovas and black hole catastrophic explosions that occur in our universe, somewhere, beyond our solar system. Now, that is the connection between chemistry and astrophysics.

How stars die by exploding in novas and huge ones in supernovas. Chemistry of the Periodic Table is also involved beyond the solar system and our galaxy, where other places, black holes explode because of an increasing amount of mass and energy from the collision of the two black holes causing an explosion.

The Periodic Table is everywhere, and do you know that atoms can measure the rotation of the Earth's speed in milliseconds (ms, a thousandth of a second) from a huge atomic clock which is connected to the TAI (International Atomic Time) using the caesium element or Cs as a symbol.

Look at the image below, it looks like some kind of saber sword or something, but it is a neon blue caesium tube, responsible and crucial for the atomic clock's inner working:

While we are exploring the Periodic Table, let's dive deep inside the inner working of an atomic clock:

An atomic clock is a different type of clock that measures time based on the frequency of atoms. It also measures the rotation of the Earth's speed.

Elements, chemical elements have different atomic numbers and atomic weights and their properties help shape our world and our everyday life.

Like Oxygen (O, atomic number 8), helps life breathe from fauna to hominids or us, humans, and even flora. People think we mostly breathe oxygen, but the truth is, we mostly breathe nitrogen. I know it doesn't make sense. But scientists investigated that nitrogen contains 78% of its gas in the atmosphere, while oxygen remains at about 21%, and other gasses remain 1%.

Physics' connection with the Periodic Table teaches us how chemistry shaped out world, and how will it shape the future of chemistry science.

CHAPTER 3.2

PHYSICS & PARTICLES

ELECTRONS, NEUTRONS, PROTONS, AND THE THEORY OF RELATIVITY

Have you ever wondered how many subatomic particles are found in the human body? There are a lot of subatomic particles inside your body, and they can move around if you create friction which causes static electricity.

The chemical symbol for electron is e^-, which means the subatomic particle has a negative electric charge, repels when it contacts with a proton, and it neutralizes when it is with a neutron.

Electrons are subatomic particles that have a negative electric charge opposing protons, Do you know what a subatomic particle is?

- A subatomic particle is a different kind of particle that is smaller than an atom, like neutrons, protons and electrons.

There is an antimatter similarity of an electron, known as the positron. The positron has the same amount of electrical charge like a normal electron, but the charge is opposing an electron, meaning its positive.

They are different forms of particles known as a quark. A quark is a different form of a particle that creates protons and electrons, based on their electric charges inside atoms.

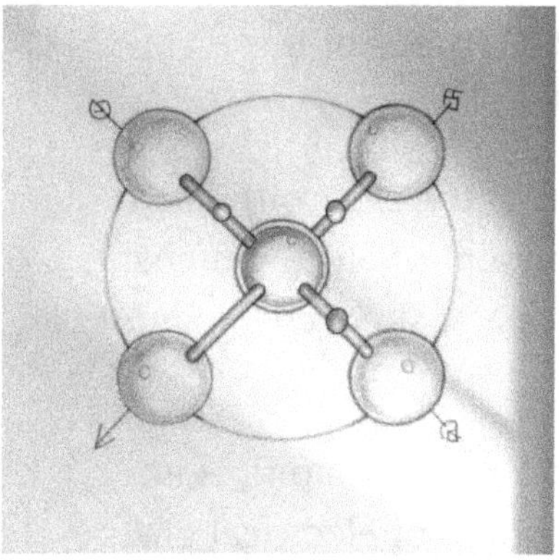

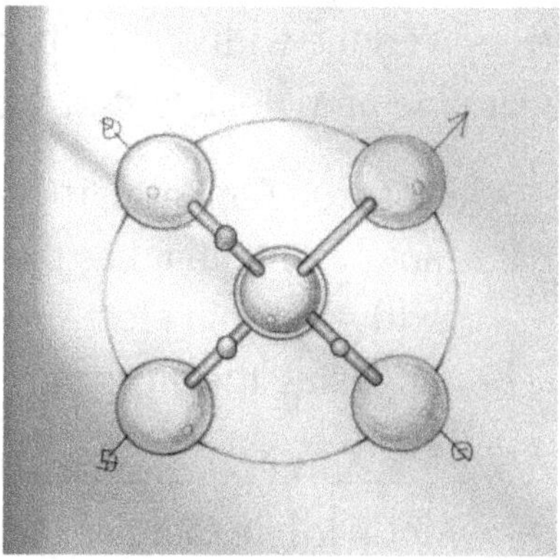

Here is the quark structure of a proton on the right, showing two quarks are down meaning more positive, and one quark is up meaning less negative.

The quark structure of a neutron on the left, showing two quarks are up meaning more positive, and one quark is down meaning less negative. But has no electric charge

The chemical symbol for proton is e^+, obviously meaning the subatomic particle has a positive charge based on its mass. While the neutron has no electric charge on its mass.

In particle physics, electrons have negative charges if they meet with other electrons, they repel, similar to magnetic physics. If protons are with electrons, they attract.

Electrons, neutrons and protons have a quantity of about spin ½ of a lepton in an atom. A lepton is a different kind of particle that carries electric charges electromagnetically, meaning these subatomic particles both carry electromagnetic interactions at a half spin of a lepton.

We aren't done with the antiparticle counterparts of these subatomic particle, lets start with:

- Positron (an antiparticle version of the electron); is an antiparticle that has the same amount of electrical charge with a normal electron, but it has a positive charge opposing the electron.
- Negatron (an antiparticle version of the proton), is an antiparticle that has the same amount of electrical charge with a normal proton, but it has a negative charge, opposing the proton.

  Neutrino (an antiparticle version of the neutron), is an antiparticle that has the same amount of electrical charge with a normal neutron, but it has a neutral charge, but no electric charge, opposing the neutron.

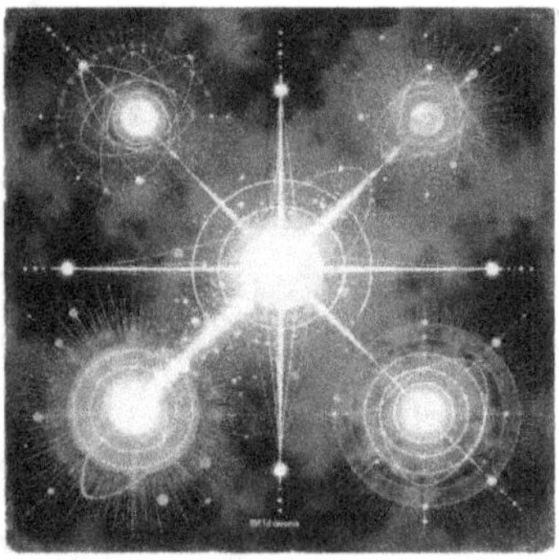

If our particles, (electrons, neutrons and protons) meet with the antiparticles (positrons, negatrons and neutrinos, it would cause a catastrophic chaos of explosions between the two counterparts.

Well, matter and antimatter don't become friends because of their oppositions. Antimatter has various amounts of antiatoms which oppose normal atoms, these two counterparts would both explode, if they contacted each other.

There is another particle known as a photon. A photon is a different kind of particle that has a quantum of light based on its mass and electromagnetic force, which it is a particle found from the Sun's ultraviolet rays.

The image above shows an expert who emits photons in a laser beam from a laser showing how much light is inside these photons emitted.

Photons are also elementary particles that are transmitted from the Sun's rays to the solar panels that are made from the element Silicon (Si, atomic number 14), which acts as a semiconductor. Photons can create photovoltaic energy, meaning it is a different kind of energy that powers the solar panel to produce electricity.

People wonder what is the difference between a molecule and a particle?

- A molecule is a structure that is made of different particles or atoms like two hydrogen atoms and one oxygen atom make H^2O or water.
- A particle is the smallest form of matter that is found in atoms moving around, creating a spin like a proton having a positive electrical charge.

In the date of 14th March 1879, the most astonishing German physicist and professor Albert Einstein was born in the land of Ulm, Kingdom of Wurttenburg in Germany. Then many years after 1879, the professor made the most popular theory, physicists are learning today, the Theory of Relativity!

Einstein's Theory of Relativity is divided into two variants of the theory:

- Einstein's General Theory of Relativity; focusing on gravity and its works.
- Einstein's Special Theory of Relativity; focusing on space, time, the speed of light, the formula $E = mc^2$, their works in our universe.

Let's start from Einstein's general theory of relativity:

What is the general theory of relativity? The general theory of relativity is a type of theory of relativity that Einstein stated in the 1900s, it focused on the inner workings of gravitation from Newton's laws of universal gravitation.

Gravitational force or gravity is a type of force that pulls or exerts two objects between each other in space based on their mass.

For example:

- The Sun has 99,85% of mass in the entire solar system, which means gravitational force pulls all planets and they all revolve around the Sun in different days or years.

Mercury is near the Sun, causing the planet to revolve faster meaning, it ends its revolution for about 64 Earth days or two Earth months to be precise.

Neptune is further away from the Sun, causing the planet to revolve slower meaning, it ends its revolution for about 164 Earth years. While our Earth takes 365 ¼ days or a year to end its revolution around the Sun, stated on Newton's laws of motion and laws of universal gravitation.

Einstein's general theory of relativity is also occurred on the gravity that the black hole have. Yes, the black hole is a form of matter that exerts strong gravitational pull towards celestial objects and can bend light from its gravitational lens.

Here is the image of the black hole releasing quasars below, taken by HST:

Black holes are an awesome example to show what gravity does in the universe. There are other black holes out there that are ten times bigger than this one on the image. Those ginormous black holes are called supermassive black holes, they are 10 to 50 times bigger than normal black holes. The black holes have an event horizon where the strong gravitational pull takes place, that exerts that other objects towards those black holes.

We are done with the general theory, now we step into the world of the special theory of relativity stated by Prof. Albert Einstein.

What is Einstein's special theory of relativity? The special theory of relativity is another type of theory of relativity that focuses on mass, energy and the speed of light's workings.

Einstein stated that matter or mass and energy can be equivalent to speed of light in a vacuum. For example: an explosion has an increasing amount of energy and mass, with light showing on that explosion. That concept is called the mass-energy equivalence or formulated as $E = mc^2$. E is energy, m for mass or matter, then c is for the speed of light in a vacuum or empty space.

Energy is equivalent to mass multiplied by the speed of light squared (c^2). Speed of light is equivalent at an accelerating rate of 299,742,458 m/s^2. So we square the speed of light rate ($299,742,458^2$) which is equivalent to 599,484,916 m/s^2 in the $E = mc^2$ formula.

Particles like photons, neutrons and electrons release an increasing amount of electromagnetic energy based on their electric charges.

Physics and particles are related to each other, through chemistry and particle physics. Einstein's theories of relativity are also related to physics based on forces and astrophysical quantities.

CHAPTER 3.3

PHYSICS & STATIC ELECTRICITY

HOW HUMANS CREATE THEIR OWN POWER

Have you ever wondered how us, humans create our own electricity, statically? Electricity, it is everywhere, across our planet, even our universe. The word electricity was made by Greeks known as "electros".

Why do we humans create our own electricity? Static electricity is a different kind of electricity that we humans create from objects that have different electric charges. Friction is the cause of why we humans create our own unexpected electricity.

When we humans rub our objects towards other objects; Those objects receive electrons from one object to another, creating static electricity. When you contact the rubbed object towards a metal, it creates a small spark. Same thing happens during storms, when clouds have different charges, similar on the ground also has different charges, causing the biggest spark on Earth, also known as lightning.

Static electricity can also be a danger in some other places like the gas station, at home, or during thunderstorms.

- An accidental spark near a gas tanker, can cause fire and explosions which may lead to death or injury at the gas station, or even at home, if electric charges are attracting one another.
- During thunderstorms, don’t run because running causes friction on the ground which may lead for the lightning to strike on that ground.
- Lightning can also strike in tall places, like the tree or the electric poles or even street lights.

Subatomic particles like electrons, can carry a negative electric charge between the clouds and the ground, creating a lightning bolt. Similar to humans, from your finger to a metal object like a metallic door handle, creating a mini-lightning bolt or an electric spark.

Let's travel beyond static electricity, and discover different kinds of electricity we have in our universe.

- Electric charges; property that is made from subatomic particles like electrons.
- Electromagnetism; different kind of magnetism that is actually produced by electricity.
- Electric field; different kind of electricity that is powered by force.
- Electric current; the movement of electric charges around an object, measured in amperes.

Electromagnetism is a different kind of electricity that is being produced on magnets, creating a magnetic field.

For example: the copper wire wrapped around a metal, the copper wire connected through the battery, on both the negative and positive terminal, and then acts as a magnet that attracts metal using the power of electricity, this experiment was first accidentally conducted by Hans Christian Oersted in the 1800s.

Hans Christian Oersted accidentally turned on the electric current towards a compass needle, which led the compass needle move, while he was doing an experiment. Physicist Michael Faraday also stated the law of electromagnetic induction. Inductance meaning that the conductor changes in electric current that creates a high amount or less amount of voltage.

From the first ever electric light bulb to the future of electric powered transport, electricity has sparked our world with lights and electronic devices. Also we humans create our own electricity from friction. Physics and electricity have a sparking connection that we humans cannot even imagine, but it can be dangerous.

CHAPTER 4

EVERYTHING YOU NEED TO KNOW ABOUT PHYSICS & MATTER: PART I

Have you ever wondered why we breathe, why we have life, from flora to fauna, and what we are made of, scientifically and biologically? Well, this is the first part of the Physics & Matter chapter in this book. You will explore how matter has literally shaped our universe. Matter is made of atoms, as we think of. But there is more to matter that we humans cannot even perceive.

Matter is the fundamental part of our universe, every planet is made of matter, every object, every us or every human, every life out there, even extraterrestrial life is made out of matter. There are three states of matter that the world only knows: solids or ice blocks, liquids or water, gases or hydrogen for example. But there is the last and fourth state of matter, that only physicists know; plasma.

You will explore what plasma is on the next sub-chapters.

Here is an image of the plasma lamp:

There are events that matter takes place somewhere in our observable universe, that other humans cannot know.

For example:

- Violent explosions like supernovas release a huge amount of energy, that is ginormous amounts of small matter like electrons and protons fighting violently, microscopically.
- Car collisions happening in our Earth, showing a unique example of how collision physics and matter interact.
- Liquid transforming into a solid, water turning into ice; showing a state of matter.
- Matter and antimatter explode when contacted in outer space, in particle physics.

These are the three states of matter: solids, liquids and gases, and lastly plasma as the fourth state of matter.

CHAPTER 4.1

PHYSICS & ASTRONOMY I

THE UNIVERSE AND ITS CELESTIAL OBJECTS

Have you ever wondered what is beyond our night sky? Well, actually, astronomy has told us the stories of the first lyre in a form of a constellation, known as Lyra. Astronomy was first taught by Ancient Greek physicist Ptolemy. Our stars in the night sky has taught stories from ancestors passed down from generation to generation. These myths are also being taught from astronomical constellations. Here's the Sirius, the brightest star in the sky.

But there's more to that. The real astronomy. Other people wonder what astronomy is? Astronomy is the science that focuses on the celestial properties on celestial matter or objects ike stars, moons, planets, galaxies and even supernovae explosions happening beyond our universe, and even a battle between dark matter and visible matter.

Beyond our Solar System, there is a circular field that is made of rocks, ice and water around space named after German astronomer Jan Hendrik Oort. This field is called the Oort Cloud, and it was discovered by him in 1932.

The coolest thing about astronomy is that there is nebulae beyond our galaxies and solar system. Nebulae is an immense cloud of gas that is also known as the "birthplace of celestial stars" in our universe.

Astronomy and physics have taught us how our universe works even every fundamental part of the physical world, like time, speed and distance. In astronomy, astronomers measure in units called light-years, light-minutes, or light-seconds as well. They measure how many light-years is a far-away celestial object far from Earth and the entire solar system.

Other people think the universe ends at the final distance of our solar system which is also Neptune. An incredible breakthrough happened in 2005, when Pluto got out of the solar system's path and started to be a dwarf planet.

Pluto was the ninth planet of the solar system. How did this all happen? It happened because of the planet's weak gravity and had a low gravitational pull towards its closest planets like Neptune and Saturn, causing the planet to be removed out of the orbit. Making Pluto one of the most known dwarf planets.

Pluto is a dwarf planet until today, since 2005. Have you ever wondered what a dwarf planet actually is? A dwarf planet is a celestial object that has cleared its path from the solar system's orbit.

Well, what do you actually see in the windows when you wake up in the morning or during the day? You see, a ball of light shining in the blue sky, guess that ball of light you see every day.

That ball of light is obviously the Sun. People think the Sun is the day, gifting people light or illumination, and they think the Moon is the night, that shines in the night sky. But actually, it is all a myth. Yes, do you know why I say it is a myth?

Yes, the Sun does shine in the day, but it is not just a huge torchlight that helps you perceive in the day. It is the source of life, that helps flora (plants) and fauna (humans and animals) to live like photosynthesis. Photosynthesis defines as the process that helps plants manufacture their own nutrients and food.

The Moon doesn't create light itself, in the night sky. The Moon actually it is the reflection of light rays from the Sun, making it look like it is shining during the night. Other people think, the Moon illuminates itself because of its reflective surface.

Back to Jan Hendrik Oort's incredible discovery, The Oort Cloud is the home of comets. Comets are different particles of ice and water that have incredible amount of energy, flying across the Oort cloud.

Let's travel through our universe and beyond, where we discover more about our universe and its matter inside it. The most phenomenal and yet astronomical systems we have on our universe, is the Solar System, which is made up of eight orbits with planets revolving around our Sun. Each containing different surface compositions, like rocky and gassy planets. There are four rocky and gassy planets out there,

- Mercury, Venus, Earth and Mars are rocky planets,

- Jupiter, Saturn and Neptune are gassy planets.

Celestial objects in the Solar System include:

- Mercury, the closest planet to the Sun and also the first planet in the solar system, has a high temperature of about 166° C, the surface is rocky, and it is the only celestial object that has no moon.
- Venus, the second planet in the solar system, has a high temperature of approximately 136° C, the surface is volcanic and very hot, the atmosphere on that planet is mostly carbon dioxide (CO^2). Venus has no moon.
- Earth, the third planet in the solar system, and has a normal temperature of about -15° C and higher and our home. The surface is full of land and sea, and the atmosphere is mostly oxygen (21%) and nitrogen (78%). It has one moon revolving around its spherical structure, but it is called the Earth's Moon, it is visible in our night sky every night.

- Mars, the fourth planet in the solar system, and has a lower temperature of approx. -21°C. Mars is astonishing because of its reddish surface. Scientists sometimes call planet Mars as the "red planet". Also has no moon like Mercury and Venus.
- Some believe there could be aliens living on that planet. Mars is the most explored planet, because scientists and businessmen like Elon Musk discover it using advanced technology like probes and space objects to discover what is happening on Mars. In 2011, Probes called Curiosity and Opportunity were sent to Mars to discover any red samples on the planet.
- Next on the list, it is the most incredible asteroid belt. The asteroid belt is the path between Mars and Jupiter discovered by scientists. The most famous dwarf planet in the inner Solar System out there, it is Ceres.

Here is the image of Ceres, the image taken by HST.

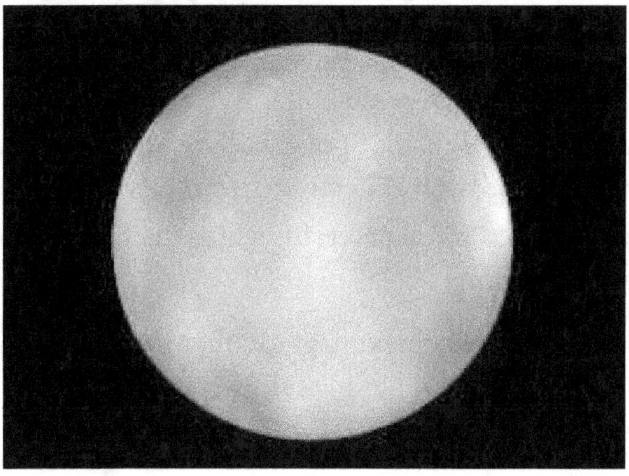

- Ceres, is an astonishing and fascinating dwarf planet that rides in the asteroid belt between the two planets, Mars and Jupiter. The asteroid belt is also known as "the separation or isolation between rocky and gassy planets".
- After the asteroid belt, it is the gas giants of the solar system, but starting with the fifth planet of the solar system, the great Jupiter.
- Jupiter is the fifth planet of the solar system, yet the first gas giant of the solar system, it is made mostly of hydrogen gas, but when you see on images, people think its surface is hard, but the surface is intangible yet gassy.
- What is so amazing about Jupiter is that, it has one of the most catastrophic anticyclonic storms ever made in the universe, also known as the giant and Great Red Spot.
- The Great Red Spot is the greatest catastrophic anticyclonic storm in the universe. It might look like a spot from afar, but near it is huge, catastrophic and uncontrollable.
- The Great Red Spot is similar to Neptune's Great Dark Spot and Small Dark Spot. But they are both anticyclonic storms (opposes real cyclones from Earth), but the Great Red Spot has more powerful winds than Earth's normal cyclonic storms.

Look at the image on the next page, Jupiter has one of the most astonishing spots in view, but in reality, it is hazardous and catastrophic:

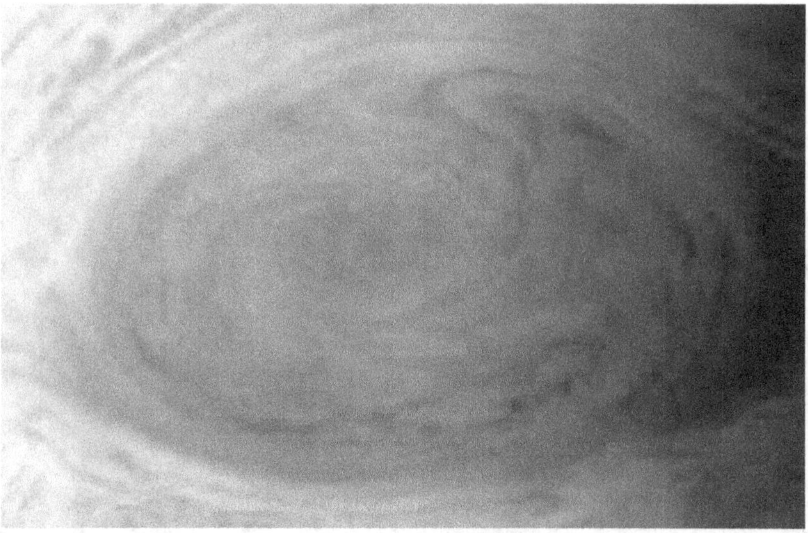

- Jupiter has four moons that are dangerous if humans were present in those moons, Io, Callisto, Europa and Ganymede.
- Io has high amounts of volcanic substances in its surface, like its other moons, Callisto, Europa and Ganymede.
- After Jupiter, we continue our journey to Saturn. Saturn is the sixth planet in the solar system, and yet the second gas giant in the solar system. Saturn is one of the most beautiful celestial objects in the observable universe, the reason behind its beauty, it is the rings covering the planet Saturn.
- The rings covered around the planet are not just rings but these rings are actually made of little particles known as rocks and ice.

The view of the planet is astonishing because of Saturn's incredible rings on the next page:

Saturn has an average radius that is nine times better than the Earth's radius.

- After Saturn, we go to Uranus, the seventh planet in the solar system, the ice giant, the coldest and blue planet and it has nine invisible rings made of rocks and ice. Saturn has 64 moons.
- Uranus is astonishing because of its bluish and frosty surface. Uranus has a lower temperature of approximately -263° C almost an absolute zero temperature. Uranus has 128 moons revolving around its sphere.
- Neptune is the eighth planet in the solar system, and terminates its revolution for about 164 Earth years. It has a lower temperature than the Absolute Zero at about -274° C. It is more colder than Uranus. It has the Great Dark

Spot, Small Dark Spot and white clouds in its atmosphere that are made of methane (CH^4) ice

- Also, Neptune has more than 128 moons, but one of the most known moons is Triton. Triton is the only moon in the universe that revolves around the planet in a weird way.

On that image shown above, it is the external planetary structure of Neptune

1. Inside the planet’s structure, The upper atmosphere showing white clouds made of methane (CH^4) gas.
2. The surface made of only blue ocean-like hydrogen, ammonia (NH^3), and methane gas. There is no hard surface, the planet is made of mostly gas.
3. The bluish planet Neptune's mantle is not just hot like our Earth's mantle but it is the coldest mantle in the solar

system. The planet's mantle is made mostly methane ice, water and hydrogen.

4. The planet's core is made of rough ice and has gas particles inside it.

Then our journey continues beyond our solar system to Pluto, the dwarf planet and the Kuiper Belt.

- Pluto was the ninth planet of the solar system, until it became a dwarf planet in 2005. It has no moon. Pluto has a very weak gravitational pull between other small celestial objects in space.
- Pluto is the second most massive dwarf planet in the Solar System, while Eris is declared as the most massive dwarf planet in the solar system.
- Eris has 27% more mass than Pluto, and has 0,27% of the Earth's average mass.

Here is the image between the two incredible massive dwarf planets in the universe, Eris and Pluto on the bottom of the page:

- The first image is the most biggest, most heaviest and most massive dwarf planets of all; Eris. This light indicates that the image was taken by the HST (Hubble Space Telescope) and NASA.
- While the second image is the second most massive dwarf planets of all, but lighter than Eris; Pluto. The surface is a bit dark and yellowish on the sides of the spherical dwarf planet.

We journey beyond our solar system to the amazing space belts of all, the place is dark; the Kuiper Belt!

Yes, the Kuiper Belt is beyond these dwarf planets and it is also the final part of our solar system. The Kuiper Belt was first discovered by astronomer Gerard Kuiper in the mid-1800s. This image shown on the bottom-right of the page, shows data gathered by NASA Science Cenre, showing the objects far away from the solar system and beyond, rocks afar from the Sun.

The Kuiper Belt is approximately 30 AU (astronomical units) to 50 AU from the Sun, extending the orbit of Neptune. An astronomical unit is equivalent to the distance that the Earth has from the Sun. The Earth's distance away from the Sun is about 150 000 000 km (one hundred and fifty million in numerical words) and 15×10^6 km (in scientific notation) which is also equivalent to an astronomical unit (1 AU).

If I calculate 50 multiplied by 150 million, it should be equivalent to 7 500 000 000 or 7 billion and 500 million

kilometers away from the Sun. The Kuiper Belt (in km distance) is about 4 500 000 000 to 7 500 000 000 kilometers away from the Sun.

There is also a dwarf planet in the Kuiper Belt, called Makemake. Makemake is the largest dwarf planet in the Kuiper Belt object where there's a lot of tiny asteroids in the belt.

The image shown above was taken by the NASA, National Aeronautics and Space Administration, showing that Makemake is a beam of light but actually it is a dwarf planet in the Kuiper Belt.

In the first part of the Physics & Astronomy chapter, we explored how celestial objects in the solar system also in the universe have different compositions and how other surfaces from other celestial objects or planets are differentiated from Earth's surface.

CHAPTER 4.2

PHYSICS & TIME I

THE CREATION OF TIME SINCE THE TIME OF THE BIG BANG

Have you ever wondered how much time did it take for the heat of the Big Bang explosion to cool down? It actually took 375 000 years for the heat to cool down.

After the cooling, it was called the afterglow light pattern, when the light made from the explosion started to dissolve, but the universe continued expanding until today, which happened 13.8 billion years ago. "That time was the time before time", that is what British physicist Brian Cox said on his ABC (Australian Broadcasting Corporation) documentary Life of a Universe.

People wonder what is time, and how it was created? Why does time affect our lives and our aging? People wonder where did all the time go? Time is not just the thing that you check all day, and ask another person, what is the time?

Time is an irreversible process where events take place from the past through the present and then to the future and its property is divided into SI base units like seconds, milliseconds, microseconds, picoseconds or more. Irreversible meaning you cannot be able to go back in time or reverse time itself. That is all fiction.

Time is also the fourth dimension found in the spacetime curvature and in Euclidean physics, found in four-dimensional shapes like a tesseract. People think time goes faster than possible in these upcoming years, but that is just theory. People think time goes faster because they have a mental disorder, they are busy with too many things everyday and unable to manage their own time, this disorder is called tachysensia.

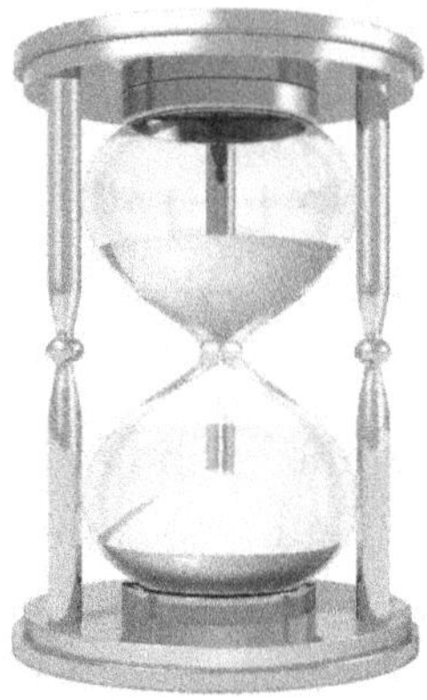

Let's go back in time since the creation of our Universe, the Big Bang, a catastrophic explosion that happened 13,8 billion years ago, before the beginning of the Dinosaur Era on Earth. How time was created? Time was created based on how many time frames did the explosion take action or explode for a specified amount of time. Those smaller units are called Planck times, named after German physicist Max Planck.

This hourglass shows how much a flow of sand is pouring down to bottom part of the hourglass, representing a second in time they say. In ancient philosophy, philosophers used to say hourglass is like money, you cannot waste that precious money or time. But you can save it all for your bright future.

The hourglass also represents as a form of three events that have happened in time, the past (historical events still remembered today), the present (recent events that are happening in our world, right now), the future (the events that will happen tomorrow and upcoming days).

Time has been with humanity for years to come. Ancient civilizations also created their own timekeeping devices like a sundial. The first ever sundial was built in the BC (Before Christ) era in some other ancient civilization, it uses a gnomon that creates a shadow underneath the sundial from the sun representing an hour from a local time.

This is a horizontal sundial from Taganrog. During the Renaissance Era, Italian polymath Leonardo da Vinci invented the first ever wrist watch to tell people the time. But other physicists like Galileo Galilei created some other new advancements in accurate timekeeping, such as a quartz watch which we have today like Rolex watch products. In the technology of time, we have quartz watches. Let's discover the internal mechanisms of how wristwatches work.

Wristwatches work based on the movement it creates and the sound it makes, like the tick-tock sound. The movement mechanism of the watch is incredible.

The watch and the clock's movement mechanism has three hands,

- The short hand which is the hour hand,
- The long hand which is the minute hand
- The thin longer hand which is the second hand.

Time is also involved with the history of the calendar. The other parts of the world mostly use the Gregorian calendar, with twelve months in a year. Pope Gregory XIII discovered the original Earth days mostly end in a revolution around the Sun for about 365,25 or 365 ¼ days. A leap year occurs every four years. A leap year has an extra day in a year which is 29 February. In other parts of the world they call 29 February leap day, in other traditions they also celebrate the day.

Time is the most fundamental part we have in our universe, imagine what will happen without time? We won't know how long is the TV show airing on screen, there would be no existence of the fourth dimension in Euclidean physics. We will also not know what happened in the past, the present and even the future.

Time can also be a fundamental part in space, when merging with the spacetime curvature found underneath planets that we cannot ever think of, based on the planet's mass and gravitational pull it is exerting from other matter. Time can never be reversed, it is irreversible like I explained earlier. Some philosophers and scientists say that when an object rotates around the Earth at a high speed, it will cause time travel, but is that true?

Well, all that theory, it is based on science fiction. Time travel is another fictional form of transportation that is being mostly used on science fiction (sci-fi) movies like Back To The Future and its incredible trilogy. Time travel is a physicist's most loved fictitious thing that is occurred in the world of time physics. Time is also found in speed. Where time is being divided by both the distance the object creates and the speed it makes.

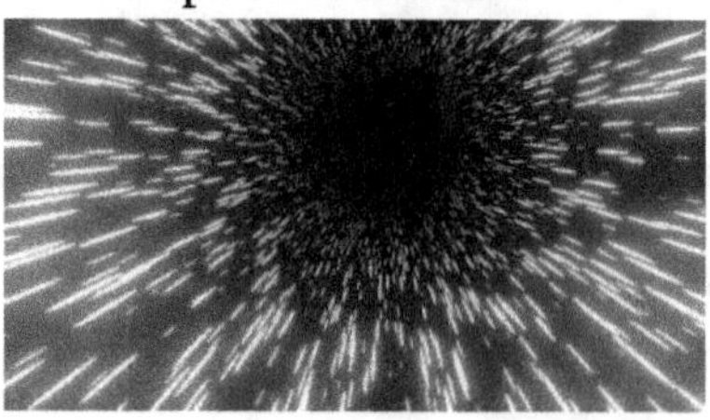

CHAPTER 4.3

PHYSICS & STATES OF MATTER I

THE FORMATION OF GAS

Have you ever wondered how gas is created? It is all the power of atmospheric and matter physics. I know, gas sometimes can be visible, but other natural gases are invisible, but while using the microscope, the gases are shown to have small molecules made of other atoms.

For those who do not know the power of matter, how is gas formed? It is an interesting question for young minds. Well, gas is formed by the power of atmospheric physics. Atmospheric physics is a type of physics that focuses on weather and its events that is causing during climates or environmental issues.

Gas is actually formed when water is being heated by the Sun's rays, causing the water to release small gas-like particles called water vapour. Then, water vapour is released into the air, making clouds. That process is called evaporation. This same process also happens in the water cycle.

Drifting smoke particles showing an example of gas movement. In a microscopic glimpse, gas can be shown as small gas particles flying around with no contact with each other, and the space is bigger than the gas particles itself,.

Gas is one of the four fundamental part in states of matter (including solids, liquids and plasma).

Gases can also be found on gas giant planets like Jupiter, Saturn, Uranus and Neptune.

Gas can be found in different types of its property like CO^2 (carbon dioxide, dry air) and O^2 (oxygen, air that plants produce from the photosynthesis process to help us humans breathe on Earth). Jupiter has hydrogen gas in the atmosphere, Uranus hasbluish gas made from the mixture of gases like hydrogen (H), ammonia (NH^3) and methane (CH^4).

There are also poisonous, hazardous and harmful gases out there that can eliminate and kill you, like Sulfur (S) atomic number 16 in the Periodic Table. Sulfur can be a corrosive and harmful element, it is also involved when making a sulfuric acid compound (H_2SO_4).

All noble gases like Neon and natural gases like methane are all flammable, and when they contact with fire, it causes an explosion. Nitrogen, also a gas can be liquified as well into liquid nitrogen.

CHAPTER 4.4

PHYSICS & GRAVITY

THE UNSEEN FORCE

Have you ever wondered what force causes all the planets to revolve around the Sun in the Solar System? It is all gravity. What is gravity? Gravity is the most fundamental part in our universe, that which leads all objects with mass attracting one another.

Gravity is also related to Newton's theory of gravitation, about all the planets revolve around the Sun in the Solar System, also the Moon revolve around the Earth's orbit. But what if there is a small moon revolving around the Moon? Is it true?

Well, scientists are gathering information about the Moon and its gravity around the Earth. Gravity can also be found on the Earth's tides, this type of gravitational force, it is called tidal gravity. These tidal forces are being exerted from the Moon and the Sun, cause tides on our Earth's oceans. What are tides?

Tides are the rising and falling of our Earth's ocean caused by the gravitational tidal forces caused by the Moon and the Sun. Here is the image above showing how tides affect the Bay of Fundy with high water (the tidal forces rise the Earth's ocean water)

The image on the bottom-right side of the page showing how tides affect the Bay of Fundy with low water (falling tidal forces in the Earth's ocean water).

The tidal physics is all happening in the Solar System, when the Moon changes direction, causing the tidal forces to change the rise or fall of the Earth's oceans.

The revolution of the Moon around Earth creates a force of gravity that causes the Moon to cause tidal forces when Earth has different seasons on each hemisphere.

When the Northern Hemisphere has autumn and summer, the tides start to fall, and when the Southern Hemisphere has spring and winter, the tides start to rise, and vice versa.

Tidal forces exerted by gravity, can cause natural disasters, like a tsunami on Earth, In theory, People sometimes call tsunamis tidal waves. Actually, tsunamis appear after the earthquakes have concluded on that place. But in reality, tsunamis aren't caused by tides.

This is an image of a tsunami caused by the gravity of the Earth taking place in the Indian Ocean by the year 2004.

Tsunamis are actually caused during post-earthquakes, and by the force of nature not tides caused by gravitational force.

There is something beyond gravity that we humans don't understand, but theoretical physicists understand, and that is supergravity. Supergravity is a type of gravitational force that is a theoretical framework that deals with supersymmetry and the spacetime curvature that was theorized on the general theory of relativity. Supersymmetry is also involved in dimensional physics.

Supersymmetry is also found in four-dimensional spaces, but physicists haven't found out yet about the existence of supersymmetry, for example; the spacetime curvature found underneath celestial bodies because of its mass and size, the spacetime curvature stated on Einstein's general theory of relativity has also four dimensions, length, width, height and time.

Gravity is also theorized and stated on Einstein's general theory of relativity inside black holes. Inside black holes, there is an event horizon. An event horizon is the void inside the black hole that exerts a stronger gravitational pull and captures matter from not escaping, and has a singularity that Professor Stephen Hawking explained. The singularity of the black hole inserts nearby objects using a every destructive gravitational pull that can crush a nearby celestial object into smithereens or bits, and then that object can never escape the mysterious black hole, also light cannot escape as well.

This unseen force in our universe known as gravity, has changed our minds, by seeing fall and rise like the tides in the ocean, and rocket falling to Earth, with fire that comes out of nowhere from our planet's atmosphere and high gravitational fall.

Gravity is not just a force, but the way of life. Gravitation is the way of life, because people do not fall when the Earth is revolving around the Sun at an average speed or velocity.

Gravity is also involved in one of Newton's fundamental laws of motion; Newton's 2nd law of motion, it states that "f = ma or f = m × a", in literal form, it states that force is equivalent to mass multiplied by acceleration. But the equation for gravity; it is w = m × g, weight is equivalent to mass multiplied by gravitational acceleration.

Here is a calculation example:

How to find the gravitational value or force in newtons? Mass is equivalent to 4,5 kg, then the Earth's position of speed or acceleration is about 9,8 or 10 m/s^2

$f = ma$

$= 4{,}5 \times 10$

= 45 N (the force is about 45 newtons, when the mass is 4 ½ kg and the Earth's acceleration is 10 m/s^2)

Gravity and physics are the fundamental concepts in our modern world, they help us understand the real world and in the virtual world.

How the Moon's gravitational force affects the Earth's tides into the highs and the lows of the ocean, and how the Earth and other planets revolves around the Sun by gravitation, and how the black hole exerts a powerful gravitational pull and crushing planets into smithereens. Gravity is not just a normal force, but a powerful one. Gravity also helps us defy the laws of physics.

CHAPTER 4.5

PHYSICS & DARK MATTER

ITS INVISIBLE SURROUNDINGS BEYOND OUR SOLAR SYSTEM

Have you ever wondered that our world is made of matter? There are different types of matter; dark matter, visible matter and how matter states like gas, liquid, plasma and solids. Visible matter is a different type of matter that we humans perceive like TVs, fridges, and other visible items we know.

But, dark matter, is course, dark matter has no color, but it is invisible beyond our Earth and the entire solar system. The solar system has visible matter, while the entire observable universe has dark matter. What is dark matter?

Well, dark matter is actually a type of matter that has almost total mass in the entire universe, and dark matter has constituted 23% of the universe's mass, and dark energy has 78% of the universe's mass and since 13,8 billion years ago during the Big Bang, subatomic particles like neutrons and protons, and atoms were constituted at a small percentage rate, while dark matter had 63% of the universe's mass, seen on the pie graph below.

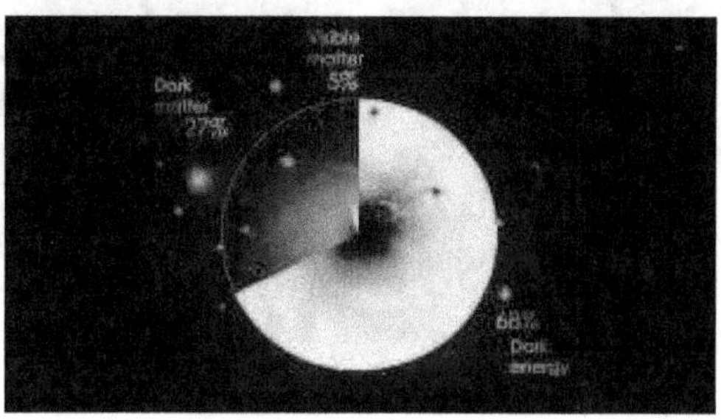

It proves that dark matter has more mass in the microscopic world such as particles since 13,8 billion years ago, but 13,8 billion years later, dark matter has less mass, while dark energy has more mass. Dark energy is made up of dark matter. There is not only dark matter in our universe, there are bits of luminous matter or visible matter like stars, gas, ice and some bits of water in outer space.

The scientific evidence was proven by Dutch astronomer Jan Hendrik Oort in 1932, about some bits of visible matter beyond our Solar System, or other scientists call it the surrounding of oursolar system, which is made of luminous matter like stars, and other tiny celestial bodies out there, it is the Oort Cloud.

This is the data collected by scientists years ago, about the Oort Cloud and the inner and outer part of the solar system, and the Sedna orbit.

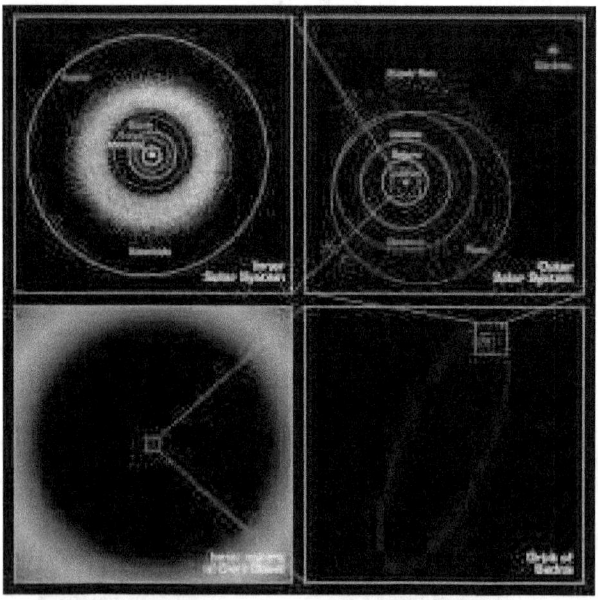

Dark matter constitutes 84% mass of the whole observable universe, while luminous or visible matter constitutes 23% mass of the whole observable universe. But the master of all this dark and luminous matter is radiation, not just any radiation you know like uranium and polonium discovered by Madame Marie Curie.

But the radiation that made our entire universe known as the Cosmic Microwave Background Radiation or CMBR. CMBR was discovered in the 1940s and it is an electromagnetic radiation that has a high amount of microwave radiation frequency at about 160,2 GHz and a super sizzling temperature of 2,725 K, hotter than the heat of our Sun. CMBR has a wavelength of 1.9 millimeters (mm) or 0,019 meters (m).

The radiation's temperature will drop when the universe continues expanding. The CMBR is also involved in the universe's expansion, the spot where the universe continues the expansion process, it is called the surface of last scattering.

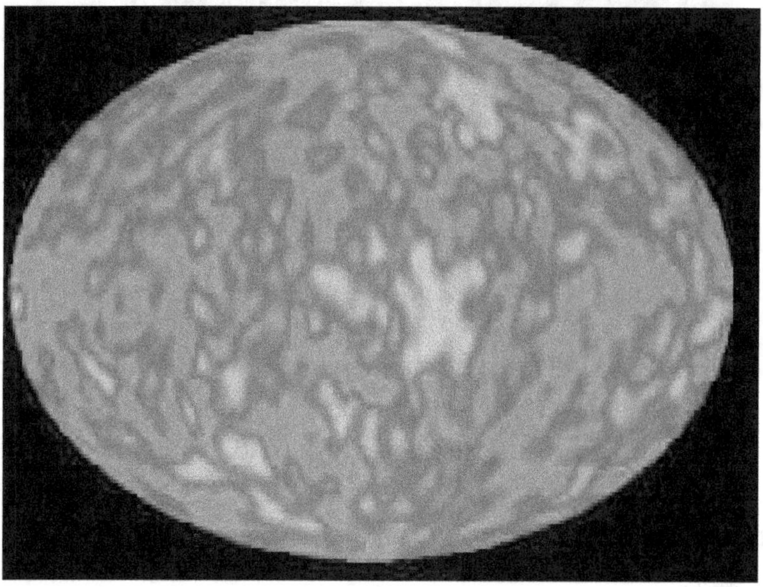

What is amazing and fascinating about dark matter?

The thing that dark matter is amazing and fascinating is that it cannot be seen through telescopes, and does not emit or absorb light from other sources, that is why it is called the "invisible matter" from other astrophysicists.

This is an image showing strong gravitational lensing shown on the Hubble Space Telescope showing the presence and existence of dark matter within those lensing arcs. Lensing meaning that there are other parts of celestial bodies and objects that emit and bend light in outer space.

Physics or astrophysics and dark matter have shaped our observable universe from the first ever explosion to the creation of our universe or the Big Bang to the understanding of how dark matter and luminous/visible matter differentiate, and how dark matter is termed as invisible because it cannot be seen on the human eye, or even on objects that help us see afar like telescopes.

CHAPTER 4.6

PHYSICS & LIGHT

COLOR'S RELATION WITH LUMINOSITY, EMR AND THE INTERACTIONS OF ENERGY AND MASS

Have you ever wondered how color has a connection with light? Well, color is shown in different varieties of its property, red, orange, and other beautiful colors. But how does color relate to our thing that enlightens or illuminates our minds, and produces electricity, and the thing that helps us see in the dark?

Let's start with white light. White light is divided into seven colors that appear on the rainbow when water droplets from the sky with light from the Sun refracts dividing the light to seven colors creating a rainbow when it is raining on the Earth's atmosphere. A rainbow acts as a spectrum of light that has 7 different colors. White light is divided into seven colors; red, orange, yellow, green, blue, indigo and violet (ROYGBIV).

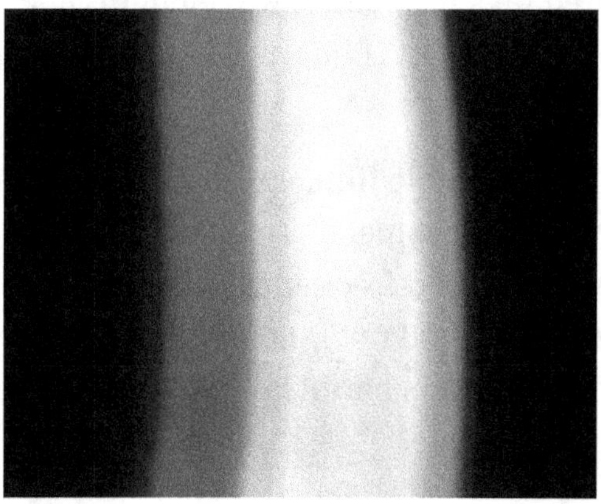

Light or visible radiation is a source of electromagnetic radiation that can be perceived by the human eye. Other people out there do not even know that the light energy they are using is made of radiation but it is harmless when using it, but can be harmful on the human eye, by causing ultimate blindness and photosensitive epilepsy. Light can be divided into seven ROYBGIV colors.

For example; a reflective surface like a triangular prism when light enters the reflective surface of the triangular prism, the light is being refracted into seven different colors from the rainbow like during precipitation and the Sun present.

Light has infrared rays has longer wavelengths and lower frequencies, and ultraviolet (UV) rays has shorter wavelengths and higher frequencies.

The electromagnetic visible light spectrum has wavelengths that start from 380 to 720 nanometers (nm) or 0,000,000,380 to 0,000,000,720 meters, and frequencies start from 350 to 420 terahertz (THz) or 350,000 to 720,000 gigahertz, that releases a lot of radiation, but light is harmless in our daily life, but it is harmful on the human eye.

Different types of EMR or electromagnetic radiation are also involved in our daily lives. For example; a television remote when surfing television channels, you cannot see infrared rays coming from the side of the remote where you point it on television,

Let's discover what electromagnetic radiation is. Electromagnetic radiation is a self-propagating or processing wave that forms in an electromagnetic field. The electromagnetic spectrum starts with gamma rays. Gamma rays are a different type of electromagnetic radiation that atomic nuclei has a radioactive decay or nuclear instability. Radioactive decay meaning, that it is a process where atomic nuclei particles become unstable to other atomic matter.

Gamma rays has shorter wavelengths and longer frequency. Gamma radiation (symbolized as ʎ) has a wavelength of more than 350 picometers (pm) or in scientific notation, it is actually 35 $\times 10^{-16}$ meters, and a longer frequency of 500 GHz to about 300 THz of radiation.

In the early 1800s, physicist Ernest Rutherford named the first ever radiation that receives an amount of radioactive decay (at that time), he named that radiation "gamma rays" because of its strong amount of frequency and higher radiation. The gamma rays are related to X-rays. But X-rays have lesser radiation than gamma rays.

Gamma rays are astonishing, they can be found in space and these radiation particles come far from Earth and the satellites around our planet receive the gamma ray particles. Why do gamma rays have shorter wavelengths but longer frequency?

The reason of gamma rays having shorter wavelengths and longer frequencies, it is because in the electromagnetic spectrum, the wavelength are inverse with frequency, meaning when the wavelength is shorter, the frequency is lower, and when the frequency is higher, the wavelength is longer, it is all opposition or inverse. Then after gamma rays, that is when the X-rays come up.X-rays are a form of electromagnetic radiation that has deals with photography and imaging, but very dangerous when it has exposure to humans. X-rays were discovered by German radiologist and physicist Wilhelm Roentgen.

During the early 1900s, X-rays were called Roentgen radiation, named after Wilhelm Roentgen himself. Roentgen also named after himself that the see-through images made by X-rays called roentgenography. X-rays are also very useful in hospitals, observing what disease or condition does the patient have, by looking at the interior part of the patient's body.

On the date, 12th of January 1896, a photographer H. L. Smith took the first ever X-ray built photograph.

Here is the modern X-ray photograph taken by other doctors, showing the internal parts of body in black and white:

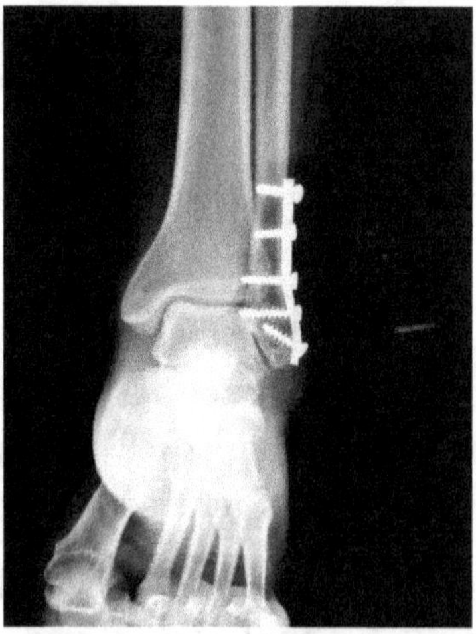

An X-ray was called an X-ray because it was identified as an unknown type of radiation as for Wilhelm Roentgen. X-rays are actually have shorter wavelengths than those of UV rays and have longer wavelengths than gamma rays and have frequencies that range from 30 petahertz to 30 exahertz (3×10^{16} Hz to 3×10^{19} Hz).

This visual symbol shows the danger that the X-ray causes due to radiation exposure:

- Radiation exposure from X-rays causes cancer inside our bodies
- It also causes radiation sickness, an illness that damages cells inside our bodies, like coughing blood and having terrible headaches
- X-rays that Wilhelm Roentgen discovered, are astonishing by identifying the interior of our human body to check bones and organs, but have a disadvantage of causing radiation exposure and cancer.

The X-rays are an amazing form of electromagnetic radiation that Prof. Wilhelm Roentgen discovered in 1896, but this kind of radiation can be deadly. It cannot be perceived to the human eye.

Wilhelm Roentgen, one of the fascinating scientists who created such an astonishing but deadly radiation that can cause cancer into the human bodies, but useful for medical observations inside our human body.

After the exhilarating X-rays, we move to the ultra-amazing ultraviolet rays or UV for short. Ultraviolet rays are a form of electromagnetic radiation that can be found in the Sun's solar flares and sunlight on earth, ultraviolet rays have shorter wavelengths than that of visible light, and longer wavelengths than X-rays, at a length of 10 to 400 nanometers (nm) or in decimal form it is about 0,00001 meters to 0,0004 meters.

Ultraviolet rays constitute 10% of the total electromagnetic radiation that the Sun exerts from outer space. But ultraviolet rays are deadly like other forms of radiation like X-rays and gamma rays. The Sun exerts a lot of UV rays from its solar flares, causing sunburn when we haven't apply sunscreen in our skin. Excessive sunburn causes skin cancer, a type of tumor that damages our skin cells. Ultraviolet rays don't only cause sunburn and skin cancer, but they can also damage our DNA inside our bodies. DNA stands for Deoxyribonucleic Acid.

Deoxyribonucleic Acid or DNA is a genetic structure that sends genetic instructions inside our bodies like transmitting cells, andit is also involved in human reproduction. DNA has also four chemical bases. Those chemical bases are referred as GCTA, which stands for Guanine, Cytosine, Thymine and Adenine. These chemical bases are part of the double-helix structure that the DNA creates.

Here is the double-helix structure that the DNA creates:

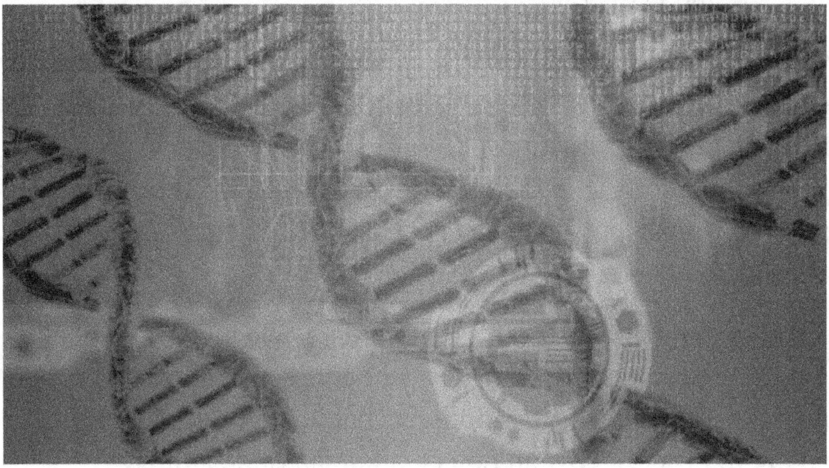

This is why it is important to apply sunscreen when you are visiting the beach and playing in the sun. Ultraviolet rays are very dangerous, but sunlight plays a huge role in plant production, from seed growth to photosynthesis. The Sun has different variants of ultraviolet light coming out from the Sun's solar flares, UVA, UVB and UVC. These different variants of long-wavelength ultraviolet rays like UVA, UVB and UVC, they are actually a form of UV radiation that enters through our human skin, inside the layers of our skin, and damaging the skin cells.

UVA rays have longer wavelengths than UVB rays at a length of 320 to 400 nanometers and can damage the inner layers of our skin, causing skin cancer and a higher risk of melanoma, and those rays can reach into the lens and retina of our eye.

UVB rays have shorter wavelengths than UVA rays at a length of less than 320 nanometers, can cause your skin to be dark and cause less amounts of skin cancer cells inside your skin. While

UVC rays cannot reach to the Earth's surface because it is blocked by the Earth's atmosphere, has the shortest wavelength than UVA and UVB rays.

Have you ever wondered why radiation is part of controlling the receiver while surfing the television channels? What I mean is that, a remote control you use to surf or change channels has a type of electromagnetic radiation inside known as IR or infrared radiation, which appears in red color in the electromagnetic spectrum.

Infrared radiation is also used in different ways in our everyday lives, like observing different kinds of temperatures on other objects, this kind of observational technology is called thermal imaging to observe how an object, a human or an animal's temperature is like.

When the object is colored from dark blue to light blue, it means the object is in a very cold temperature from -30° C to 10° C, measured on a thermal imaging camera. When the object is colored gteen, the temperature is normal. When the object is colored red to yellow, it shows that the object is in a high temperature and produces more thermal energy or heat energy.

Infrared radiation can also be used to perceive cosmic objects like cosmic dust or gas from telescopes used by astronomers and physicists. Physics and light radiation have an enlightening relation that help us "illuminate" our thirst for knowledge and help us perceive everyday, everywhere you go and every time.

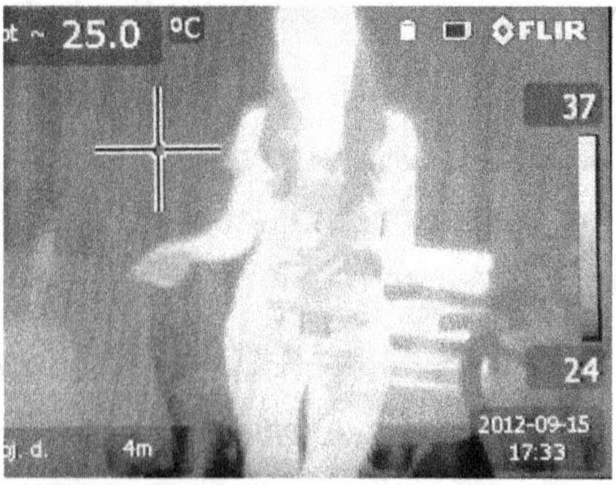

CHAPTER 4.7

PHYSICS & WAVES

HOW INVISIBLE RADIO BROADCASTING AND WI-FI INTERACTS WITH WAVES

Have you ever wondered how every radio device is airing the same broadcast simultaneously? Well, obviously, the radio device you are tuning is being broadcast through invisible radiation from a single radio station where they broadcast.

Exactly, the radio programmes that are you are tuning to, is being broadcast using radio waves from satellites from space like IS20 (Intelsat 20) which is mostly used in South Africa to broadcast both television and radio programs on different receivers and decoders, while other Intelsat versions are used in other areas around the world. What are radio waves?

Radio waves are a form of electromagnetic radiation that have the longest wavelengths and shorter frequencies in the electromagnetic spectrum. Space satellites and satellite dishes from earth transmit radio waves from space to our radio devices we use to tune in to our radio devices. Radio waves can also be transmitted on our devices we use everyday like smartphones from mobile data, 5G (Fifth Generation Long-Term Evolution) to Wi-Fi.

Wireless Fidelity (or Wi-Fi for short) use the 2.4 GHz to 5 GHz radio wave bands to send or receive data which allows any kind of device to connect to the internet without Ethernet or any kind of cables that are used to connect to the internet.

Why does Wi-Fi use radio waves for? Well, Wireless Fidelity uses radio waves for, data transmission, reception from devices that connect to the internet and decoding radio signals into internet data.

- Wi-Fi uses radio waves to convert radio signals into internet data
- These radio waves which work with its known frequency bands or signal strengths like 2.4 GHz or 120 mm in wavelength UHF (Ultra High Frequency) and 5 GHz or the half of 120 mm which is 60 mm in wavelength SHF (Super High Frequency) are being transmitted into the air from the router's antenna.
- Our Wi-Fi=connected devices are receiving internet data from the radio waves the router is transmitting
- Wi-Fi also converts back radio signals into digital information from the internet

What do you think will happen when there will be no Wi-Fi and radio broadcasting? This is why we think of disadvantages that might happen in the real world. What will happen if there was no internet, no mobile data, no TVs, and no technology? Some people will say, it will be the end of the world. Obviously, maybe, there will be a protest for bringing Wi-Fi back, ot there will be World War III, which is also be called the War of Bringing Back Technology, by killing those who are responsible for Wi-Fi and other types of technology, and for those who listen to radio and watch television, they will miss their favorite programs and get disappointed for those who broadcast channels around the world or nationwide. Maybe, those disappointed people will burn those satellite dishes for revenge and hate. Okay, let's get back to the topic.

So, there are many radio broadcasting companies in South Africa and around the world who use electromagnetic radiation physics and the power of satellite transmission to entertain us with their satisfying programs, like South African Broadcasting Corporation or SABC for short, Primedia, Kagiso Media, Kasie FM somewhere in Sam Ntuli Mall, Katlehong, in the South African province of Gauteng and other radio broadcasting companies out there who provide radio programming for their listeners everyday.

In South Africa and other parts of the world, we listen to radio from the VHF signal strength or FM broadcast band that starts from 87.5 MHz to 108 MHz.

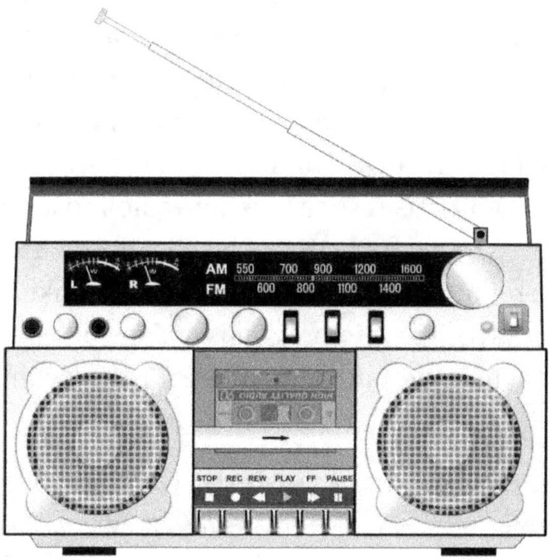

These two images here shows an antenna receiver that receives radio waves from the satellite in outer space, it is the most useful object that helps broadcasters provide both television and radio programming for listeners, and another image shows a radio cassette which is mostly used by old people to listen to their favorite modern radio programs, the most fascinating thing is that, it has a built-in antennae, so you can listen to radio clearly without any disturbances or any rough noise the radio causes when the antenna is far away from the radio.

The 87.5 to 108 MHz band is also called the VHF Band II (Roman numeral for 2). The VHF Band II is also part of the ITU Region 2 somewhere in Europe and in Africa as well for listeners who listen to radio every day. What is the ITU? ITU or International Telecommunications Union is the agency that is owned by the United Nations and is responsible for operating information and coordinated technologies internationally.

The ITU also allows the radio spectrum and satellite orbits set technical standards for interconnected networks. The ITU was established 160 years ago in 1865, and it plays an important role to make sure that people have the right to communicate with their family when they are away from them and have access to technology. The ITU is responsible for radio broadcast bands like VHF Band II (87.5 to 108 MHz) which is mostly used in Europe, somewhere around the globe and in Africa, to make sure people listen to their favorite radio programs and keep them entertained, and also listen to important news of what is happening around our planet and so on.

Radio waves can be useful in our everyday lives, but can also dangerous when exposed to humans and animals. So what is the danger behind radio waves? Inside microwaves, Wi-Fi routers at home, satellite dishes and decoders, there is huge amounts of radiation energy that is being exposed causing heat. But that heat is not just normal heat like the fire produces. It is the heat caused by excessive amounts of RF radiation, causing radiation sickness, cancer, can damage our internal body tissues and can cause cataracts meaning that it heats up our eye lens until it is no longer stable, meaning you won't be able to see anything but suffer from the radiation exposure caused by excessive RF radiation, coming from unstable devices.

Here are some safety tips of avoiding excessive RF radiation or radio waves in your area:

- Materials like copper and aluminium foil can absorb or block radio waves
- Reduce the excessive amounts of time you use your smartphone too much and connecting to Wi-Fi all the time, it boosts RF transmission power
- Last tip, increase the distance far away from the Wi-Fi router and use aluminium foil as your shield against excessive RF radiation exposure

Since this sub-chapter is all about how both radio and television broadcasting helps viewers and listeners to keep themselves entertained through the power of radio waves and satellite orbits itself. Let's explore into the electromagnetic spectrum, not about light radiation this time, but about modulation or radio frequency (RF) radiation and its bands.

Yes, RF radiation is one of the most useful radiation that helps broadcasters provide radio programming for listeners, and it can also be dangerous, we learnt that. The electromagnetic spectrum is focusing not only light, like we learnt before, but focuses on radio frequencies like UHF and SHF.

The electromagnetic spectrum has eleven different types of frequency which are measured in kilohertz, megahertz or gigahertz and different wavelengths which are measured in kilometers, meters, or centimeters.

These eleven different types of frequencies are what broadcasters use to broadcast their radio shows to listeners, but highly dangerous when near high-level radio satellites, which transmits high amounts of RF radiation in the air. Here, starting from extremely low to extremely high:

- The Low Frequencies

- ELF – Extremely Low Frequency: ELF is a form of electromagnetic radiation that is generated by electrical systems and devices like smartphones and appliances, and it has a frequency of about 1 Hz to 300 Hz which is extremely low indeed, but the disadvantages of ELF that it can cause irritating effects on your nervous system
- SLF – Super Low Frequency: SLF is a form of electromagnetic radiation that is between 30 and 300 Hz,

and it is used for military communication inside the ocean when travelling on deep-sea submarines because SLF can penetrate seawater, meaning it can be transmitted in the ocean, also in the air as well.

- ULF – Ultra Low Frequency: ULF is a form of electromagnetic radiation that is between 300 Hz to 3 kHz and its waves are found in the Earth's magnetosphere, near the exosphere. ULF helps by impacting dangerous radiation particles from entering the Earth's surface.
- VLF – Very Low Frequency: VLF is a form of electromagnetic radiation that is between 3 and 30 kHz and its waves are useful for communicating emergency situations with miners and other people who work underground by producing signals by the Earth. VLF waves are also useful by interacting with high-energy radiation particles from outer space and protect satellites from excessive radiation.
- LF – Low Frequency: LF is a form of electromagnetic radiation that is between 1 Hz and 100 kHz. Its LF waves interacts with the body's electric fields by inducing currents from the magnetic field, and it does cause cardiovascular issues and burns when exposed.
- MF – Medium Frequency: MF is a form of electromagnetic radiation that is between 300 kHz and 3 MHz. MF waves are also called medium waves or MW which are used for amplitude modulation or AM radio broadcasting and radio navigation systems like LORAN or Long Range Navigation which is a type of radio navigation system that was in the Second World War (1939-1945).

- HF – High Frequency: HF is a form of electromagnetic radiation that is between 3 and 30 megahertz (MHz), and its HF waves carry energy and information, and at high levels, it causes poor blood flow in the eyes, and damage the tissues and other internal organs.
- VHF – Very High Frequency: VHF is a form of electromagnetic radiation band that is between 30 and 300 megahertz (MHz), and it is used for FM radio (frequency modulation) broadcasting, and aviation communication. VHF is also part of the ITU Region 2 where it is mostly used for broadcasting those who use 87.5 to 108 MHz radio. VHF has shorter wavelengths and wider bandwidth which allows devices to boost data rates.
- UHF – Ultra High Frequency: UHF is a form of electromagnetic radiation that is between 300 MHz and 3 GHz, and has very short wavelengths of about 1 meter to 120 millimeters, which is in descending order. UHF waves are also used in communication systems and surveillance.
- SHF – Super High Frequency: SHF is a form of electromagnetic radiation that is between 3 and 30 GHz, it is involved in the microwave spectrum, meaning it cannot be seen by the human eye, but it can be detected elsewhere around us, like inside microwave ovens. People use specialized equipment like antennas to detect whether SHF is near or not.
- EHF – Extremely High Frequency: EHF is a form of electromagnetic radiation that is between 30 and 300 GHz, it has the shortest wavelength in the electromagnetic spectrum, they are devices that has EHF radiation like

cellphones, microwaves and Wi-Fi networks. It ends up to 300 GHz, highest frequency, shortest wavelength.

So what is the difference between wavelength and frequency? Wavelength is the length that the wave makes, creating different kinds of patterns, but it is measured in meters and it is represented by a Greek symbol known as lambda (λ).

Frequency is the number of times a repeating cycle that occurs over one unit of time, it is likely measured in hertz, which is also one cycle per second. The difference is that wavelengths and frequencies are inversely related, meaning that when frequencies increase, the wavelengths decrease, or higher frequencies means shorter wavelengths, and vice-versa.

Let's discover the formulae of frequency and wavelengths:

Formula for frequency: $f = v/\lambda$ (v for wave speed divided by a wavelength is equivalent to frequency)

For example:

The speed of the wave is about 10 m/s, then the wavelength is about 1000 meters or 1 kilometers.

$F = v/\lambda$

$f = 10 / 1000$

$f = 0{,}01$ Hz

Formula for wavelength: $\lambda = c / f$ (c for speed of light which is equivalent to 299,742,458 or 300,000,000 m/s)

For example: The frequency is actually 300 Hz when calculating the unknown number of the wavelength

$\lambda = c / f$

$\lambda = 300{,}000{,}000 / 300$

$\lambda = 100$ km or 100,000 m

We've learnt different types of frequencies in the electromagnetic spectrum and why they are involved in our planet, and how they work, and its harmful effects with the environment and our human bodies, radiation might be dangerous but helpful.

Radiation in the electromagnetic spectrum is also helpful by blocking high-energy radiation particles that come from outer space into the Earth's magnetosphere and radiation belts, and how it helps miners and underground workers who need emergency by communication, and keeping viewers and listeners entertained from the power of satellites, radio waves and all of radiation.

CHAPTER 5

EVERYTHING YOU NEED TO KNOW ABOUT PHYSICS & THE PAST

Have you ever wondered how physics travelled through time and space, from the past until today? Well, in these incredible stories of the Physics & The Past chapter, we travel back in time, and learn how physics was involved and how it played an important role in the past, and how it plays an important role today.

Discover how Italian chemist Ascanio Sobrero invented the boom-bastic chemical behind dynamite by accident, and how Alfred Nobel recreated into something "safe" which is used until today, and the discovery of pulse-stellar radio sources by a graduate student and also a woman who fell in love with science.

The past's connection with physics is very important, if there was no physics, because today, there would be no dynamite, no atomic bombs, no Wi-Fi, no television/radio broadcasting, no Periodic Table, no atmosphere, and no universe, and no particles, atoms or even dark matter, or anything that is made by the power of physics in the past. Discover at how these minds came up with brilliant ideas, that they didn't even know that their brilliant ideas would change the world someday, except Jocelyn Bell Burnell.

CHAPTER 5.1

PHYSICS & THE INVENTION OF ELECTRICITY

FRANKLIN, TESLA, FARADAY AND EDISON, THE MASTERMINDS

Let's travel 319 years back in time to the date of 17th January 1706, the electrifying polymath mastermind Benjamin Franklin was born. Benjamin was the 10th son of the 17 children of the man who manufactured soap and candles with his bare hands.

Years later, Benjamin Franklin became a polymath, a person who has numerous careers and handles them at once, like Leonardo da Vinci. Benjamin was the writer of *The Autobiography of Benjamin Franklin (which was published in 1771)*, a scientist, a political philosopher, an inventor, a statesman, a diplomat, a printer and a publisher as well. Franklin conducted a kite experiment to prove that lightning is electrical to other people, where he used a kite and a key during a thunderstorm.

Franklin used the kite and the key during a thunderstorm to conduct his experiment and proven true, the kite experiment works when a kite is tied with a conductive wire to collect some electrons in the air and conducting it into the wet kite string to the ground. It is an experiment focusing on how static electricity works.

The kite experiment was first conducted in the year 1752 by Benjamin Franklin. Benjamin was the Founding Father of the United States, he was the drafter of the Declaration of Independence in the United States of America. Benjamin also invented the most electrifying object that helps us not being struck by lightning, the lightning rod.

Nowadays, the lightning rod protects buildings and people inside from being struck by lightning. The lightning rod is actually a device that is connected at the top of the building to prevent lightning strikes during a thunderstorm.

This incredible and shockingly amazing device that Franklin invented, protects people and buildings from a hazardous lightning strike. What will happen if Franklin didn't invent the lightning rod? Obviously, there would be an increasing death toll of people who are being struck by lightning, and an increasing number of buildings collapsing and infrastructure damage.

Actually, back in time, Benjamin Franklin invented the lightning rod, not as a safety device, but as a creation to prove that lightning is electricity.

But, Franklin, the mastermind behind the kite experiment and the lightning rod, and the man who is in the dollar bill of America, didn't know that his lightning rod invention would turn out to be a safety device for preventing a building and protecting people inside the building from being struck by lightning.

We travel forward in time from the Franklin era to the Tesla era, 269 years back in 10th of July, 1856, the inventor of the Tesla coil and the plasma ball was born in the city of Smiljan, Croatia, and his name was Nikola Tesla. His parents were Duka Tesla and the most amazing priest of Croatia, Milutin Tesla.

Nikola Tesla was a Serbian-American (a cross-culture between a Serbian and an American) engineer, futurist and an inventor. Yes, in 1891, Tesla invented the first ever Tesla coil, named after himself.

So, how does this shockingly astounding device work? Well, the Tesla coil works by producing high voltages of electricity pushing electrons into the air, causing a big lightning-like phenomena.

Don’ts of the Tesla coil:

- When having contact with the Tesla coil, it will electrocute you so badly to death.
- When near the Tesla coil, it will cause very serious burns called RF burns. RF burns are not just normal burns, they are very serious burns that heat the internal tissues of your body.
- When having a Tesla coil at home, don’t switch it on near children, when they are wet, electricity contacts with water.

The Tesla coil is an electrical resonant transformer invented by Nikola Tesla in 1891, 234 years ago, and one of the most shocking, electrifying, and sparkingly amazing electrical devices ever invented.

Let's learn what else Nikola invented, the plasma ball, invented in the year 1894, 231 years ago.

The plasma ball is a fascinating device and a glass container that is filled with noble gases, with a combination of three elements, Ne, Kr and Xe (neon, krypton and xenon).

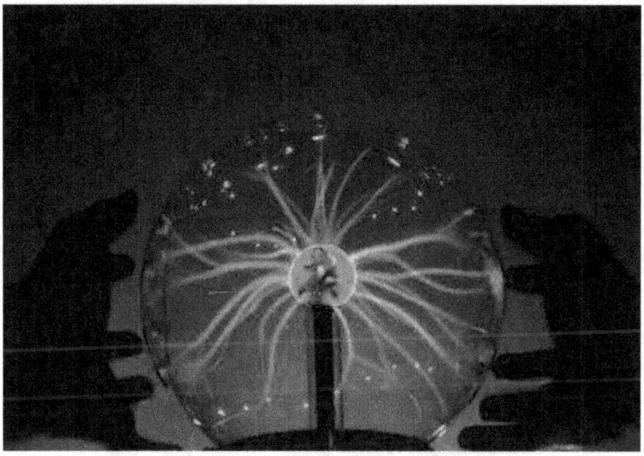

This image shows a globe with some kind of nucleus inside and its releases amounts of plasma energy made by a mixture of those three elements inside it, that is the fascinating plasma ball invented by the greatest inventor, Tesla himself.

Here is another invention that Nikola Tesla has made, it helps humans to control every device by their own command, and inside that control, it has amounts of IR. Guess what that invention is?

If you haven't or have estimated it yet, Nikola Tesla invented the first ever remote-controlled device in the year 1898, but how did it become the most useful device on earth today? But, first lets head back in time, and see how the first ever remote was like. Nikola Tesla invented the remote control because he experimented as a wider use of alternating current (AC) and had a AC vs. DC battle with Thomas Edison. It was the battle of the two currents: alternating versus direct.

The first remote-controlled device he built was an example of how his wireless communication invention worked. So he invented the remote-controlled device wirelessly using infrared rays. These images below shows Tesla's most ingenious invention of the first ever remote controlled device, and Nikola Tesla himself.

Nikola, the mastermind behind the Tesla coil, plasma ball, and the first ever remote-controlled device, has thought of brilliant ideas that he did not even know that his gadgets would make history.

We are traveling back in time from Tesla era to the Faraday era years before Tesla was born. We travel 334 years back to 22nd of September, 1791, the inventor of the Faraday cage and the electric motor was born on this day. His name was Michael Faraday, the most shockingly incredible physicist of all time.

Michael Faraday was a Danish chemist and physicist who was known for his works on electricity and magnetism as well. In 1836, Faraday invented the first ever Faraday cage named after himself. The Faraday cage works as a shield by blocking other electromagnetic waves or fields.

Well, in the world, we are surrounded by electromagnetic radiation like infrared, microwaves, X-rays, gamma rays and so on. So, Faraday built it as a protector against electromagnetic fields. Here is the image of the first ever Faraday cage invented in 1836, looks modern.

You may wonder how does the Faraday cage function? Faraday cages function when they have an equal and opposite electrical force when exposed to another electrical force that cancels the electrical force inside the Faraday cage, which leads to the cage blocking other electromagnetic fields. The Faraday cage shields the interior from electromagnetic radiation to a large degree.

The Faraday cage is not the only invention that blocks electromagnetic waves, they are materials you can also use at home, like metals such as copper, aluminium, steel, silver and nickel. These metals are mostly used to help shield electromagnetic radiation because of their conductivity especially copper. So, the Faraday cage is not only used to block electromagnetic waves, but metals that conduct like copper.

Faraday also invented the first ever electric motor in the year 1821. The first-ever electric motor looked like a current-carrying wire, meaning the wire had enough electric current or the switch was on. It looked like a current-carrying wire and a permanent magnet that created continuous rotary motion.

How did Faraday's electric motor work? The electric motor that Faraday invented works when the electric current flowing through the wire creating a circular magnetic field that interacted with the magnet, causing the wire to create rotary motion itself or rotating. Faraday's swirling but shocking electric motor, has changed and revolutionized the world of electrical engineering and physics today.

We travel 10 years after the electric motor invention, to 1831, where the inventor Michael Faraday invents the first ever electric generator that led to the development of transformers and motors today. But, what was the reason Faraday invented the first electric generator? Well, the reason Faraday invented it, because he discovered electromagnetic induction. What is electromagnetic induction?

Electromagnetic induction is the principle that a changing magnetic field that can induce or creating an electric current in a wire. So he thought of experimenting electromagnetic induction another way by inventing the Faraday disk or the first ever electromagnetic generator. Here is the image below showing the Faraday disk invented back in 1831.

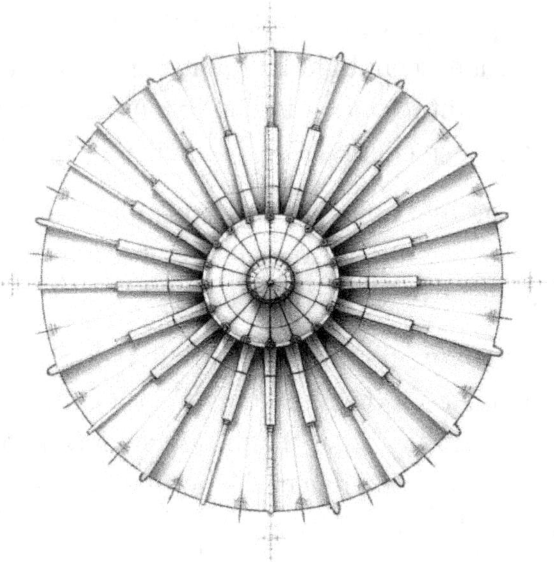

How did the Faraday disk work? The Faraday disk works when he used a magnet to create a changing magnetic field that induces or creates an electric current in a wire, now that is what I call an electricity generation, what I mean is, since Faraday's time, generating electricity has been evolving from generation to generation, until today.

From power stations to transformers that generate electricity on different areas, to provide electricity for other people who need it for cooking and doing other stuff, like watching television, or writing a book on a computer. The Faraday disk has changed the world with the help of the most "brightest" masterminds in history, Faraday himself.

We travel back in time from the Faraday era to the Edison era, where Tesla was also present. We travel 278 years back to 11th of February, 1847. The inventor of the light bulb, the phonograph and the kinetoscope himself, Thomas Alva Edison was born.

Later, at the date of 31st of December, 1879, Thomas Edison invented the incandescent light bulb, meaning that the light bulb he built emitted light because of its high temperature, releasing huge amounts of heat energy. How did Edison's light bulb work?

The incandescent light bulb works when a thin, high-resistance carbonized filament heats up to a high temperature and emits light, that is how incandescence processes.

You may wonder why didn't Edison's light bulb filament release excessive heat and burn the place down? Inside the glass bulb, it was filled with inert gas, a gas that is not chemically active, theinert gas' work was to prevent the hot filament to contact with oxygen and burn the place down.

The incandescent light bulb, one of the most "illuminating" inventions in the world that changed the world, and produce other kinds of light like LED bulbs and energy-saving bulbs, which are mostly used today. Here is the image of the first-ever incandescent light bulb invented by Edison in 1879 to 1880, and his brainstorming idea of the invention.

We travel 3 years back before the invention of the incandescent light bulb, in 1877, Edison invented the phonograph in his Menlo Park Lab. The phonograph that Edison invented that was capable of recording and replaying from the function of a tinfoil-wrapped cylinder, when Edison said “Mary had a little lamb” when replaying on the phonograph. The most fascinating thing about the phonograph is that it records any sound, it was the fascinating thing when Edison invented for other people, who were fascinated and curious of how does the phonograph do that. Here is the image of Edison’s first sound recorder.

We travel again 9 years after the invention of the incandescent light bulb, in 1889, at the year where Edison and his assistant W. K. L. Dickson invented the first ever Kinetoscope, a device that shows celluloid strips that produce motion pictures for a single film. This was the first invention that led to the generation of cinema creation. Edison, the mastermind behind the bulb that illuminates us all today and the kinetoscope that leads to the generation of cinema creation shaped the world of both illumination and creating imaginations to life. Here are the images of the phonograph and the Kinetoscope below.

So, the masterminds who invented all things electricity had some "enlightening" ideas, they have sparked our curiosities of why are transformers generating so much electricity? Who made the first electric motor? Who made the first ever remote control? The masterminds who invented all shocks and sparks that have changed our modern world, and our Earth is filled with electricity, thanks to those masterminds.

- Benjamin Franklin (1706-1790) conducted the kite experiment to prove that lightning is also electrical and invented the lightning rod, the device that prevents buildings from dangerous lightning strikes.
- Nikola Tesla (1856-1943) invented the Tesla coil, plasma ball, and the first ever remote control.
- Michael Faraday (1791-1867) invented the Faraday cage, a protective device that shields against electromagnetic fields, and invented the electric motor and the Faraday disk that uses electromagnetic induction to generate electricity, it is mostly used today on modern generators, and finally,
- Thomas Edison (1847-1934) invented the incandescent light bulb, Kinetoscope and the first ever sound recorder. These electrifying masterminds have invented some shocking and sparking inventions.

CHAPTER 5.2

PHYSICS & THE DISCOVERY OF PULSARS

JOCELYN BELL BURNELL MADE HISTORY

We travel 58 years back in time to 1967, where a young lady who fell in love with science and a postgraduate student named Jocelyn Bell Burnell, was at a site where they are radio telescopes and satellites, she observed different radio pulses from space from the new radio telescope she helped construct to study quasars.

Her observation and analysis of the data she gathered showed repeated signals from an astronomical object in space, where a neutron star spinning rapidly emitting pulse-stellar radio beams from space.

How did she discover pulsars from space? She helped construct the radio telescope when she was a PhD student at Cambridge, then later, she saw weird signals while analyzing on the telescope's data. That pattern she observed showing rapid radio pulses that appeared at every 1.3 seconds.

She kept observing the same weird pattern that shows very rapid radio pulses that appeared at the same time interval. The discovery that she has found is nicknamed as "LGM-1" which might represent "little green men". She analyzed and investigated these repeated signals, and later found out that there is a new astronomical object in space, a neutron star emitting pulse-stellar radio beams also known as pulsars.

You may wonder what are pulsars also known as pulse-stellar radio beams? Pulsars are rapidly spinning neutron stars that emit high-energy radiation, and appears as pulses of light and radio waves to an Earth-based observer like a radio telescope showing pulse signals. Pulsars form when massive stars collapse during a supernova explosion. How do these pulsars work?

The pulsar's rapid rotation causes its immensely powerful magnetic field to sweep beams of light and electromagnetic radiation into outer space. As the pulsar rotates at a rapid rate, it sweeps its beams of light, and when it passes to Earth, astronomers observe as if it is a regular pulse of radiation from space.

Pulsars spin amazingly fast, from its rotation rate of between milliseconds to other types of seconds. Pulsars are a type of star that died in the supernova explosion, representing an endpoint of stellar evolution for other massive stars.

This is the image showing Jocelyn Bell Burnell observing the data showing repeated signals of a neutron star from space spinning at a faster rate, emitting pulsars.

But, after Bell Burnell's discovery of pulsars, her assistant Antony Hewish, was fascinated of her discovery and analysis of pulse signals being repeated from space. But, you may also wonder did Jocelyn Bell Burnell receive the Nobel Prize? Actually, no. Her assistant, Antony Hewish received the No-Bell Prize, that is a pun for Nobel.

It seems unfair that Jocelyn Bell didn't receive the prize because of her gender and as a graduate student. In the 1960s, women were treated as if they are only wives who worked for their husbands, and cleaning the house, and washing the dishes, and so on, it was a form of gender exploitation or inequality, and that ishow Jocelyn Bell felt about it, but she didn't care, she was only proud of what she discovered and continuing her research projects until today.

The discoverer of pulsars herself, Jocelyn Bell Burnell became the president of the Royal Astronomical Society from 2002 to 2004, and president for the Institute of Physics from 2008 to 2010. Now in the present, in 2025, she was appointed as the Order of the Companion of Honours, as a reward for her incredible achievements she has done in the past.

Here is the image below showing the discoverer of the pulsar also known as PSM B1919+21, Jocelyn Bell Burnell and Antony Hewish.

CHAPTER 5.3

PHYSICS & THE ATOMIC THEORY

DALTON, THOMSON, RUTHERFORD, THOMSON, BOHR AND CHADWICK'S TIMELINES

Have you ever wondered who are the people who first discovered the building blocks of our universe also known as atoms, and they worked? Physicists John Dalton, Ernest Rutherford and Niels Bohr, and other physicists.

Let's travel back to Ancient Greeks' timeline, in the BC era, there was a Greek philosopher who knew the atomic theory at the earliest, Democritus. Philosopher Democritus didn't conduct a science experiment but used his philosophical thoughts. He thought that they are some kind of matter that are divided infinitely.

He knew if these particles were true, it would lead to a contradiction meaning a combination of statements, or actions that are so opposite that would confuse the person who is difficult to understand. Democritus knew that that small particles can't be divided any further because of its size.

So, Democritus named those small indivisible (unable to divide) particles "atomos" which is a Greek word for atom. Democritus was the first person to discover atoms in a thoughtful way.

There is no proof that atoms existed in ancient times, but Democritus came up with the atomic theory as an idea to prove that atoms that he thought about existed or not. But there are some ingenious physicists who wanted to prove Democritus' theory true in the 19th century.

Let's travel to Sir John Dalton's timeline, in 1803, the first ever person in the 19th century to experiment of proving that atoms exist or not. Dalton proposed his atomic theory at that time, he knew that there is matter full of indivisible atoms.

Here are the key basics in Dalton's atomic theory:

- Atoms are indivisible, he also states that matter is made by atoms which are particles that are both indivisible and indestructible.
- Atoms of elements are identical, he states that atoms of a given element like Hydrogen, have the same mass and same properties.
- Atoms of diverse elements are different, atoms of diverse elements have diverse sizes, masses and properties.
- Atoms combine in fixed whole number ratios to form compounds.
- In a chemical reactions, atoms are not created or destroyed, but they separate and rearrange into different substances.

Dalton developed an idea of atoms having different masses which are called atomic weights which led to the idea that contributed to the development of the periodic table.

So, what is the main evidence that Dalton found in his atomic theory? The main evidence he found is that all matter are made of indivisible atoms, and atoms of the same element have the same mass, same size and same properties. Another evidence Dalton found is that all atoms from different elements have different mass, different size and different properties as well. But that is not part of the full atomic theory, Dalton proposed, but other physicists took over his idea, years after Dalton's discovery of the early atomic theory, and want to know more about the structure of the atom, and what it has, like British physicist J. J. Thomson.

Let's travel to J. J. Thomson's timeline, in 1897, where he discovered more about the atomic theory he proposed, where he discovered the first ever negatively charged electron he saw, when he was busy with CRT (cathode ray tube) experiments.

The electron he discovered led to the development of his atomic model, known as the "plum pudding model". The discovery he demonstrated proved that the atom is divisible and is all made up of subatomic particles (particles that smaller than the atom).

J. J. proposed his atomic model, or the plum pudding model, to explain how the atom works and how it is structured, electrons that carry a negative charge are represented inside a positively charged sphere.

Why did J. J. call his atomic model the plum pudding model?, because subatomic particles like electrons are smaller than the atom, they stay inside the positively charged sphere itself, similar to plums in a pudding.

Here is the image of British physicist J. J. Thomson's plum pudding model.

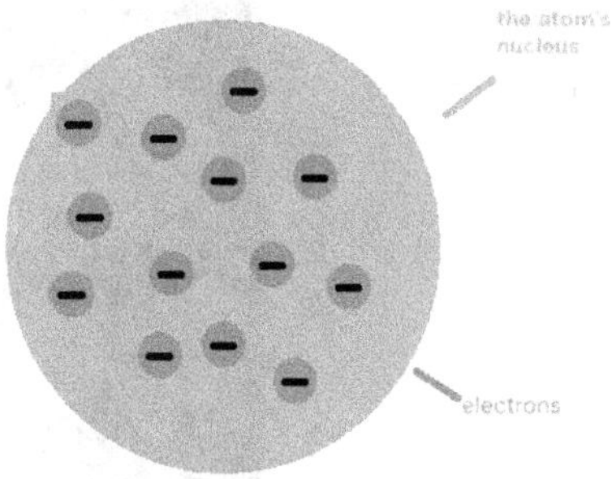

So, what is the main evidence did J. J. found in his plum pudding model and his electron discovery?

- The main evidence behind J. J.'s electron discovery because of his cathode ray tube experiments, when he realized that atoms have negatively charged particles called electrons.
- So, he used plums and pudding to propose the plum pudding model to prove that negatively charged electrons are placed inside the positively charged sphere similar to plums in a pudding, or pepperoni in a pizza, or strawberries in a yogurt.
- The plums act as the negatively charged electrons, and the positively charged sphere it is the pudding itself.

J. J. didn't name the negatively charged particles electrons, he called them "corpuscles", before the physicist who called them electrons. The plum pudding model, he proposed, was an important step and realized that atoms were not indivisible but they contained subatomic particles.

Here is the image of Joseph John Thomson and his cathode ray tube which were turned out to be negatively charged particles and the reason behind the electron discovery.

Let's travel again to Rutherford's timeline, to the year 1911, where New Zealand-born British physicist Ernest Rutherford conducted the gold foil experiment, and took over Thomson's atomic theory, and he proposed his atomic theory, and found out more about the structure of the atom.

Rutherford also proposed that the atom has a tiny, dense positively charged nucleus that is at its center containing the atom's whole mass, with negatively charged electrons orbiting around the atom's nucleus, which developed a new atomic theory.

How did Rutherford discover that the atom has a nucleus in its center? Rutherford and his assistants, Geiger and Marsden, placed a sheet of gold foil with positively charged alpha particles (particles that has two protons and two neutrons bound together). He saw the alpha particles passing through the foil, showing the atom's true empty nature.

He noticed the alpha particles changing directions at large angles, and some other particles bouncing back, proving that the atom has a positively charged nucleus in its center.

Ernest Rutherford's model replaced J. J. Thomson's plum pudding model, proving that electrons are spread out from positively charged particles, causing a force of repulsion.

What is the main evidence behind Rutherford's discovery of the atom's nucleus being at the center of its mass?

- Rutherford initially conducted the gold foil experiment to determine the structure of the atom and its properties.
- When the positively charged alpha particles change direction at large angles and some bouncing back, he

realized how the atom really works, that it has a nucleus that its center where electrons orbit around it.

- He wanted to test Thomson's plum pudding model, but the model didn't quite expected, as the alpha particles deflected or changed directions, and the positive charges were meant to spread out but they were concentrated.

Here is the image of Rutherford's gold foil experiment to prove whether Thomson's atomic theory was right or not.

The gold foil experiment shows a radioactive source releasing a beam of positively charged alpha particles traveling through the gold foil undeflected, then some other alpha particles are deflected by smaller or larger angles.

Then, an alpha particle bounces back from the foil back to the source. It proved that the atom didn't have any empty space, but had a dense positively charged nucleus which is placed in the center of the atom, repelling the positively charged alpha particles, and others deflecting and others bouncing back.

So, Rutherford's gold foil experiment was another atomic theory that had important evidence of the atom having a dense positively charged sphere inside it, which is the nucleus in its central region.

Let's depart the Rutherford timeline and travel to Bohr's timeline, to the year 1913, where Danish physicist Niels Bohr, proposed the atomic theory, where he experimented on his own model. The Bohr model.

The Bohr model works when positively charged protons and neutrally charged neutrons are held tightly in the nucleus of an atom. His idea demonstrated that electrons orbit a nucleus in shells or energy levels, not orbiting in a normal way. Electrons can only jump between these energy levels by absorbing or emitting some packet of energy, which is the quantum theory.

An electron can move to a higher energy orbit by absorbing energy which is a quantum of energy or fall to a lower energy orbit by emitting or releasing energy in a form of a photon or a quantum of light. When orbiting, electrons only exist on these energy levels. Atoms are mostly stable when their outer electron orbits are mostly full.

Electrons jump at higher energy levels, meaning they absorb energy (from heat or light), and when they fall to lower energy levels, they emit or release energy, it is all related to atomic spectra that Bohr stated.

Atomic spectra are different patterns of electromagnetic radiation like light that are being absorbed or emitted by atoms. When atoms absorb and jump on a higher energy level, it shows dark lines on a light background of the atomic spectrum, and when atoms emit and fall on a lower energy level, it shows bright lines on a dark background of the atomic spectrum.

You may wonder how does atomic spectra form and work? Well, atomic spectra forms when electrons in an atom can only exist on different energy levels, no matter how low or high the level is.

How does the atomic spectra form and work?

- Electrons in an atom can only exist on different energy levels, no matter how low or high the level is.
- When an atom absorbs a photon of light with the energy that exactly matches the difference of between two energy levels, the electron jumps to a higher energy state.
- When an electron is in a higher energy state, it falls back to lower energy level, when it emits a photon of light, which its energy is equal to the energy difference between the two energy levels, it means that it shows dark lines on a spectral and colorful background of the absorption spectrum.
- This emitted light contains of photons of light energy, which are similar to wavelengths of colors, creating spectral and bright lines on the element's dark background of the emission spectrum.

Here is the image of the atomic spectra, and the difference of absorption and emission spectrum will be shown.

Now we know how the atomic spectra work, now, let's discover the most important evidence that Bohr found about the work of electrons in his atomic theory and his Bohr model.

- The most important evidence that Bohr has proven here with his Bohr model in his atomic theory, is actually electrons not orbiting normally, but orbiting through quantized energy levels, no matter how low or high the level is.
- Electrons can only move on different energy levels (low or high), and atoms are stable when outer electron orbits are full.
- Electrons move on higher energy levels, they absorb energy, or fall to a lower energy levels, they emit or release energy or photons which are a quantum of light particles.
- Bohr's discovery and improvement of his atomic theory might also be related to how atomic spectra works. Atomic spectra works the similar way electrons do, when electrons on a higher energy level, they absorb energy, like dark lines appearing on a bright and colorful background of the spectrum.
- But, when electrons fall to a lower energy level, they emit or release energy, like bright lines appearing on a darker background of the spectrum.

Danish physicist Niels Bohr has discovered some amazing improvements of the atomic theory, and how his Bohr model works, with electrons around the shells of the atom, existing on orbiting only different energy levels, absorbing or emitting energy.

Let's depart once again from the Bohr timeline to the Chadwick timeline, in 1932, the year where British physicist James Chadwick discovers the neutron, a neutrally charged particle inside the atom's nucleus. How did he discover the neutron?

It all started, when Chadwick bombarded beryllium with alpha particles, like Rutherford did with the gold foil. Then, it produced high-energy radiation, that when those alpha particles strike a kerosene block, releasing protons, and then Chadwick analyzed the energy of the protons that were released, and knew that there is some unknown radiation that has some neutral particles with the mass similar to a proton.

In James Chadwick's astounding discovery, he completed the atomic model, by adding the neutron inside the atom's nucleus, along with protons, which is the most known atomic model until today. The model that Chadwick built, has been taught on schools around the world, and modern scientists are still researching about it today.

In his discovery of neutrons, he also explained about isotopes, isotopes are atomic masses that have same numbers of both protons and electrons, but have different numbers of neutrons, in a radioactive form of a chemical element like uranium or polonium.

The problem in previous models, before Chadwick, is that the atomic nucleus only contained protons, the discovery of the neutron was the most important missing puzzle piece of the atom's nucleus, which was previously known as the Rutherford model, without the neutron before.

Then, what is the important evidence we have found on Chadwick's discovery of neutrons?

- The evidence starts where Chadwick bombards beryllium with alpha particles, while it was directed through the kerosene block, releasing protons.
- He analyzed in the energy of released protons, that there is some strange unknown radiation of the neutral particles, with the similar mass of a proton.
- He completed the atomic model, by adding the neutron as the final subatomic particle of the atom's nucleus.
- After his discovery of the neutron, he also explained isotopes, which are atomic masses that has same numbers of both protons and electrons, and different numbers of neutrons.

Physics' connection with the past has taught us the development of the atomic theory, a scientific theory that has been proving how atoms work, and their properties, from the first early atomic philosophy of Democritus, to Dalton's indivisible atomic theory, to Thomson's plum pudding model, to Rutherford's electron discovery model, to Bohr's quantum model, and then to Chadwick's neutron discovery in 1932. These incredibly amazing physicists have taught the new generation of how atoms were improved in their timelines of discovery.

CHAPTER 5.4

PHYSICS & THE FIRST MAN ON THE MOON

ARMSTRONG'S FIRST STEP, AND ONE GIANT LEAP FOR MANKIND

Have you ever wondered how Neil Armstrong travelled in space, became the first man on the Moon? Others say, it is a lie and it is all fiction, he used some of SFX, which stands for special effects in full. Well, he used physics to travel to the Moon. Here's how it all started.

We travel back in time to the date of 16th of July, 1969, when American astronaut Neil Armstrong and his fellow astronauts Buzz Aldrin, and Michael Collins launch the Apollo 11 mission, four days before the step on the Moon. This is the journey to the Moon, but also a journey of how these astronauts, especially Armstrong used the power of physics to travel to the Moon, and his success of becoming the first man to step on the Moon.

Armstrong and his fellow astronauts, Aldrin and Collins, used the physics of Newtonian mechanics. Newtonian mechanics is built on three core principles that are related on Newton's three laws of motion.

They used a Saturn V rocket, a rocket that was involved in the Apollo 11 mission, to propel the rocket with immense hot gas downwards, with equal and opposite upward force, which is thrust, that pushed the rocket away from the Earth's ground.

This is Newton's third law of motion, which states, when there's a reaction (chemical reaction that the rocket propels with immense hot gas), there is an equal and opposite reaction. Armstrong, Aldrin and Collins also used orbital mechanics, which is a type of mechanics that uses Newton's laws of motion and gravitation to calculate and estimate the orbits of planets and spacecraft, like satellites, rockets and other space debris out there.

They also used inertia, when once free of Earth's powerful gravity, the rocket will keep moving at its intended trajectory, making less continuous thrust. Inertia is involved in Newton's second law of motion, when an object stays at rest, tends to stay at rest, and when an object stays at motion, tends to stay in motion. The Saturn V rocket had to generate more enough thrust to overcome Earth's most powerful gravitational pull it exerts and reach the Earth's orbit in outer space.

As I explained before, the astronauts used some orbital mechanics for the rocket to land the Moon safe and sound. The Apollo spacecraft or the Saturn V rocket entered the Earth's orbit, then after that, they performed a Trans-Lunar Injection or TLI for short, which is their final engine burn of the rocket, propelling it quickly out of the Earth's orbit to the Moon's surface.

The TLI burn lasted 347 seconds or 5 minutes and 47 seconds for the rocket to escape Earth's most powerful gravitational pull, and finally landed the Moon. The trans-lunar injection was the most helpful tool that helped the astronauts to achieve their objective, landing on the Moon. The astronauts, Neil Armstrong, Buzz Aldrin and Michael Collins, finally did it, they have finally made the Apollo 11 project a success. On this day, 20th of July, 1969, Astronaut Neil Armstrong first stepped on the Moon, and said "One first step for man, and one giant leap for mankind".

CHAPTER 5.5

PHYSICS & THE TRUE INVENTION OF DYNAMITE

FROM SOBRERO'S BIG BOOM ACCIDENT TO NOBEL'S RECREATION OF THE BOOM

Have you ever wondered why do we have dynamite in this world and it is used for only killing people? Some say, who ever invented dynamite, must be a treacherous person who wants to kill all humankind, well, that is not true. Dynamite was invented accidentally. Here's how.

We travel once again back in time to the year, 1846, when Italian chemist Ascanio Sobrero was experimenting in his lab, when he combined glycerol with a mixture of nitric acids (NHO_3) and sulfuric acids (H_2SO_4), then kaboom, an explosion took place at the lab, it is an accidental discovery and invention as well, the invention is called nitroglycerin, a mixture of glycerol and nitric and sulfuric acids, which have a very violent chemical reaction.

But, sadly, Ascanio Sobrero was badly injured from the explosion, and he was scared of the nitroglycerin's instability, and it might used safely as an explosive and when touching it, it may violently explode again. So, he kept this discovery as a secret, because of his serious facial injuries from the accidental explosion.

The actual cause of his serious injuries from the explosion, when nitroglycerin he invented, violently reacted, it broke all glass, and sent glass fragments everywhere, and those pieces injured Sobrero's face and hands, developing scars on his face.

20 years later after the accidental discovery of nitroglycerin to the year, in 1867, Swedish chemist Alfred Nobel, the creator of the Nobel Prize, used nitroglycerin and stabilized it. Nobel developed methods of finding a way to safely use nitroglycerin for detonation. He mixed the nitroglycerin liquid with kieselguhr. You may wonder what kieselguhr is? Kieselguhr or diatomite, is a soft, white highly porous silica-based sedimentary rock made of fossilized diatoms, which are microscopic aquatic organisms, that obviously live in the sea. Kieselguhr is mostly made of silica or silicon dioxide (SiO_2). It is also astonishing, because it can also be used as a thermal and electrical insulator. But, how did the kieselguhr stabilize the nitroglycerin?

Well, the kieselguhr stabilized the nitroglycerin because of its ability to absorb and trap the unstable nitroglycerin liquid and make it stable. Bravo for Nobel, he recreated nitroglycerin into some explosive that other people can use to detonate safely, boom comes the dynamite!, from the inventor Nobel himself. He achieved Sobrero's objective of stabilizing nitroglycerin and safest way to touch the dynamite and use it for detonating.

Nobel didn't invent the dynamite to kill mankind, but to bring peace for mankind, and show that they have the power to change the world, and that is why he established the Nobel Prize in his final will in 1895, and established it to fund the prizes for those who have achieved in physics, mathematics, art, biology, literature and other subjects for their ingenuity and creative thinking. The prizes that have been funded and awarded started since 1901, 124 years ago. Here is the image of the Nobel Prize at the bottom-right of the page.

CHAPTER 6

EVERYTHING YOU NEED TO KNOW ABOUT PHYSICS & MATTER: PART II

Have you ever wondered why you can't escape and stuck inside a black hole forever? Well, in this second part of the Physics & Matter chapter in this book, we discover more about matter, as I explained before on the previous part of the chapter. Matter is the most fundamental part of the universe, and it is made of atoms, as we think of. Matter is the thing that we humans cannot even perceive.

We will also discover more about the four states of matter and how they actually work, like three natural states of matter, solids, liquids and gases, and one powerful state of matter, that is mostly found in electricity, the only plasma. Matter is not just solids, liquids, gases and plasma, but matter is everywhere in the world, but we will also discover a type of dark matter that exerts a powerful gravitational pull on the next page, that sucks any matter near around it, matter or even radiation like light cannot escape that dark matter, it is the black hole itself.

CHAPTER 6.1

PHYSICS & BLACK HOLES

THE MORE IT GROWS, THE GREATER THE GRAVITATIONAL PULL IT EXERTS

Black holes, they are the most frightening matter in the universe, they exert extremely strong and powerful gravitational pulls near any celestial objects around it, not even light can escape its event horizon or the point of no return. But, in this sub-chapter, we discover the different amounts of matter or masses these black holes have, and how they exert different gravitational force and traps nearby objects. How do these black holes form?

Black holes form when massive stars start to decease and collapse, becoming a singularity, a point of zero volume and infinite density. It actually means, that when a star larger than our Sun deceases, it exhausts its nuclear fuel, the core of the star collapses, and the outer layers of the star explode in a violent supernova. If the remaining core is massive, the core continues to form into a massive black hole.

You may also wonder what the event horizon or the point of no return is? When nearby matter is being trapped inside the black hole's powerful gravitational pull, the event horizon is not the black hole's surface, but the boundary where those objects cannot escape from the black hole. The black hole's singularity can crush the matter that the black hole trapped.

But there is more to a black hole than meets the eye, they are other massive black holes out there, that exert highly extreme gravitational pulls from their event horizon, and other massive black holes that are found in the center of most galaxies, which also plays an important role in galaxy formation.

Also, other black holes out there are turned out to be invisible, because they absorb all light, and light cannot even escape the black hole's point of no return. Here is another question you may also wonder: how do black holes grow from normal-sized to supermassive black holes or SMBHs?

Here's how black holes grow with strongly developing gravitational pull it exerts:

- Black holes are surrounded by gas and dust and mergers with other black holes. Normal-sized black holes which are 50 times larger than the mass of our sun, can form from the destruction of massive stars, in regions where stars form.

- Large amounts of gas and dust could directly form a massive black hole of up to a thousand solar masses, meaning it is a thousand times than the mass of our sun.

- A seed black hole can increase its mass or size by swallowing all gas and dust and stars that very near, this the accretion process.

- Black holes can grow very larger when merging with other black holes.

- Supermassive black holes stay at the center of galaxies, where there are large, dense amounts of stars, gases and dust.

- These dense environments of stars, gas, and dust can also cause the seed black holes to grow rapidly through accretion and repeating mergers, where black holes merge with other black holes again and again to make it more larger, and produce greater and highly extreme gravitational force to exert near celestial objects.

- Supermassive black holes also grow rapidly through accretion and mergers all over again and again, increasing to millions or billions of solar masses, or a mass that is million or billion times greater than our sun's mass.

How do rapidly growing black holes intensify and increase their extreme gravity? Well, we all know that black holes grow rapidly by accreting matter that are surrounded around the black holes and merge with other black holes. But, as matter falls into a black hole, it forms an accretion disk where the gravity increases its intensity, pulling the matter inside until it crosses the event horizon, or the point of no return.

This accretion process, and continuously merging with other black holes, increases the black hole's mass, which means it causes the event horizon's expansion, and the intensity increase of its extreme gravitational pulls, which is the most incredible part in the black hole's rapid cycle growth.

You may also wonder, how do accretion disks form? Gas, dust and other celestial debris are being pulled by the black hole's powerful and immense gravity, these debris don't fall straight into the black hole, but into a swirling, flattened structure called, an accretion disk. That is how accretion disk form, they form when matter falls into the black hole.

Inside the accretion disk, the material slowly spirals inside as the disk loses energy, which increases the immense gravitational pull when more materials are near the black hole. Then, the matter finally reaches the black hole's point of no returns, and is consumed, it is the accretion process.

There, is also another way, black holes grow rapidly, when merging with other black holes. When two black holes collide, they merge into one large black hole, increasing its mass and gravity. With an even more larger event horizon and increased mass, the black hole's gravitational pull is more stronger than ever.

When matter is consumed by a black hole, it gets ripped apart and enters the singularity, where every matter gets crushed. But, how does black hole crushing process work? The process works, when matter approaches near the black hole, it releases extreme tidal forces, like the Moon's tidal gravity on the Earth's oceans, the matter has a stretching effect called spaghettification, which acts like a spaghetti, which turns the matter into a more weak material.

As it enters the event horizon, the matter is being torn apart into smithereens or tiny subatomic components having the same mass as other subatomic particles like neutrons and protons. These bitsare then squeezed together to form a singularity, a point of infinite gravity and density (mass per volume).

Here is the image of the insides of the black hole, the most astonishing point has infinite gravitational force and density, the singularity itself.

Why are supermassive black holes at the centers of our galaxies? What makes them part of a galaxy formation? Well, supermassive black holes have every intense and immense gravity, that they have sunk into the center of the galaxy to due to its highly extreme gravity, like a heavy rock settles into the bottom of the pond. The pond is like the galaxy, and the heavy rock is like the supermassive black hole itself. What also makes part of galaxy formation?, it is because supermassive black holes are the nucleus of the galaxy, and they are mostly involved during galaxy collisions, when black holes grow and merge larger, and their extreme gravity intensifies more and more, until it is highly immense. We, humans live on our amazing planet, Earth, because it is full of life, and Earth, is an amazing and incredible gift from the Almighty God. But, Earth and the solar system we live in, are also part of the galaxy, which is called the Milky Way Galaxy.

The center of the Milky Way, there is a supermassive black hole out there consuming gas and release high amounts of energy, powering quasars. Don't worry, the center of the Milky Way Galaxy is far away from Earth, at about 27,000 light years away to be exact. Black holes, no matter how small they are or how supermassive they are, they do exert immense gravitational pull, and matter falls into the black hole forming an accretion disk, and crushing into bits inside the singularity.

It is the way of life in the universe. Black holes might be scary and frightening for other people out there, but they are incredibly astonishing when seeing them on images like this one at the bottom-right of the page, a supermassive black hole releasing high amounts of energy, and powering quasi-stellar radio beams as well.

CHAPTER 6.2

PHYSICS & ASTRONOMY II

BEYOND OUR SOLAR SYSTEM, MORE CELESTIAL OBJECTS AND OUR MILKY WAY GALAXY

Have you ever wondered what is there beyond our Solar System, and reason why Halley's Comet travels every 76 years when visible from Earth? Well, in this second part of the Physics & Astronomy chapter, in this book, we discover more beyond our astronomical surroundings like our solar system, and beyond our astronomy, to explore astrological myths that are true or fictional, and how they are connected to horoscopes.

What is there beyond our solar system? Well, beyond our solar system, they are extra planets that orbit a star, those planets are called exoplanets. They are planets orbiting a star similar to our planets orbiting the Sun. Scientists discover that there rocky planets (similar mass to Earth), and super-Earth planet (a planet massive than our Earth, but smaller than Neptune), and ice and gas giants (one is similar to Jupiter, but other gas giant and ice giant planets are more larger than our solar system giants).

Exoplanets are also called extrasolar planets, because they are planets found beyond our solar system, and they orbit on other stars. Here are some three exoplanets out there beyond the solar system.

KELT-9b: It is the most extreme and hottest planet located approximately 670 light-years away from Earth, it is the hottest ever known planet at an extreme temperature of 4,300 degrees Celsius, that elements on the planet would melt, like water and carbon dioxide would be torn apart. It is known as the ultra-hot Jupiter exoplanet, but it is 2.8 times more massive than Jupiter's mass.

The KELT-9b planet orbits its own star KELT-9, for about 1 ½ Earth days or 36 hours, a very short period for star's constant proximity or presence.

The planets also orbits on another star, which is a B-type star called the HD 192689, it is a very hot, and highly luminous star. KELT-9b was discovered by Kilodegree Extremely Little Telescope or KELT for short in 2017.

KELT-9b's extreme heat can cause its atmosphere to expand and become less dense, which makes the atmosphere weak. The incredibly extreme radiation from the KELT-9 star can break apart molecules from the planet's atmosphere, such as water, carbon dioxide and hydrogen.

On the dayside of the planet, the hydrogen molecules are shredded up into atoms, and elements like iron exist as gas, forming an iron sky on the exoplanet. On the nightside, the planet cools down its temperature. In 2022, KELT-9B is hottest exoplanet, at a temperature hotter than our Sun, reaching at a incredibly high temperature of 4,300 deg Celsius.

We move to the second exoplanet that is beyond our solar system, it is one of the most fascinating exoplanets in the history of astrophysics, the exoplanet that is luminous on its own, even though it has a reflective surface made of silicates and metals in its planetary surface, making the planet so bright and luminous. It is the LTT 9779 b or Cuancoá.

The LTT9779b planet is a Neptune-like exoplanet that orbits around a G-type star, a star that is like our sun, called the LTT9779, or the Uúba. So what is so fascinating about the exoplanet? The most fascinating thing about this luminous exoplanet is that it is also an ultra-hot Neptune-like planet, it is similar to Neptune, but it is more times massive than Neptune, and the exoplanet has metallic clouds that reflects 80% of the LTT9779 star's luminosity, which shields the atmosphere from being blown away by the star's intense luminosity.

LTT9779b exoplanet can also orbit close to the star, with a year lasting less than one Earth day or 0,8 Earth days. There is also a region called the Neptune desert, meaning that any planet that has a Neptune-like size, cannot survive there orbiting around the LTT9779 star which is also a G-type star, those kind of Neptune-like planets will lose the atmosphere due to the star's intense heat.

The exoplanet acts as a giant mirror orbiting around the star, meaning that the exoplanet has an astonishingly high reflectivity, its metallic clouds shield the planet's atmosphere, reflecting up to 80% of the star's luminosity, making the exoplanet one of the largest known most fascinating "mirrors" in the whole universe.

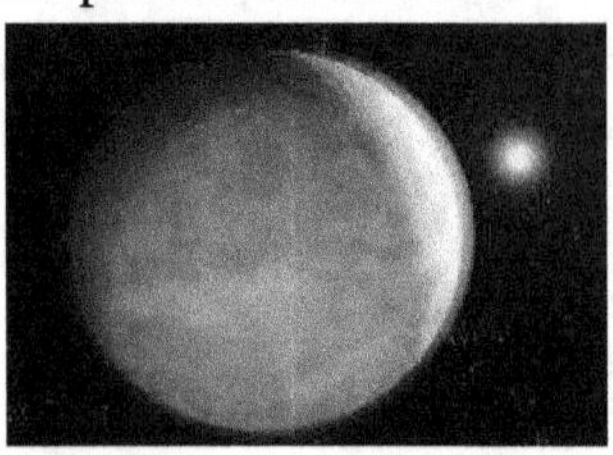

Here is the image of the most brightest, shiniest and most luminous exoplanets in the universe, LTT9779b

We move to the third and final exoplanet that is way beyond our solar system, the most Earth-like exoplanet you've ever seen before, but it is very super. The Gliese 667 Cc, it is a super-Earth exoplanet, discovered 13 years back in 2012. that orbits into the red dwarf star, the Gliese 667 C's habitable zone.

The Gliese 667 Cc, is approximately 23.6 light years from the Earth, and it is 3.8 times more massive than the Earth's size, It completes its orbit around the Gliese 667 C's habitable zone for 28.1 Earth days almost an Earth month. The exoplanet receives 90% of the light Earth receives from the Sun, but the light Gliese 667 Cc receives, it is not visible light radiation, but IR or infrared radiation, which creates the possibility for liquid water on the exoplanet, maybe they could be some possible extraterrestrial life on the exoplanet. The Gliese 667 Cc, is also a member of the Gliese 667 triple star system.

What is the Gliese 667 triple star system, the Gliese 667 triple star system is a type of star system that lies in the Scorpius constellation at approximately 23 light years away from our planet Earth. Since the Gliese 667 Cc is a member of the Gliese 667 triple star system, Earth is also a member of the Alpha Centauri triple star system.

The Gliese 667 triple star system has three stars, Gliese 667 A, Gliese 667 B and Gliese 667 C. Gliese 667 A, it is a K-type orange dwarf star, which is 73% mass of our Sun.

While Gliese 667 B, has a different story, it is the same K-type orange dwarf star like Gliese 667 A, but it has a smaller percentage of 69% mass of our Sun. Gliese 667 A and B both orbit each other at an average separation of 13 times the distance between the Earth and the Sun or 13 AU (astronomical units).

The Gliese 667 C, is the only red dwarf star in the Gliese 667 triple star system, that has 33% mass of our Sun, It orbits the A and B pair at a distance of 230 astronomical units.

When Gliese 667 Cc, is too close to the Gliese 667 C star, it could be tidally locked, meaning that the first hemisphere of the planet would face the star, which is the dayside with extreme high temperatures, and the other hemisphere of the planet faces away, which is the nightside with cool temperatures. The Gliese 667 Cc doesn't have an axis that tilts at an angle of 23,5 degrees, and no seasonal changes like Earth has.

The Gliese 667 Cc does have an axis but that tilts at angle of zero degrees, the axis is perfectly straight and perpendicular. It also has a terminator zone where both temperatures can be moderate for liquid water to exist on the planet. The planet's rotational period on its axis can be equal, as the planet has the same face on its star, synchronizing their rotation. Since the Gliese 667 Cc exoplanet, has a tidally locked axis, causing the planet to stay the same face, like day stays day forever, the temperature rises and keeps on going, and night keeps staying as night forever, the temperature cools down extremely, and the temperature keeps on cooling until ice grows on that planet, possibly.

Here is the image of the super-Earth exoplanet that could support life, meaning maybe liquid water might exist on the Gliese 667 Cc, orbiting on the red dwarf star, Gliese 667 C's habitable zone.

But, let's go on a ride more beyond our solar system all the way to our galaxy, which is the Milky Way Galaxy, where our Earth and our solar system are involved inside the galaxy.

The Milky Way Galaxy is our home spiral galaxy filled with thousands of stars, and 100 or 400 billion of gas, dust, stars and other matter, with the supermassive black hole at its center. The Milky Way is also the largest galaxy in the Local Group, which is part of the Virgo Supercluster, then the Virgo Supercluster, is also the part of the largest structure known as the Laniakea Supercluster. The disk of the Milky Way is 100,000 light years across. The Sun of our solar system is also located at the spiral arm of our Milky Way, about halfway from the supermassive black hole center of our galaxy. How do we see the Milky Way from Earth? Well, we perceive the Milky Way, as faint, and hazy band of light in the night sky, that's how it got its Milky Way name from. The stars are like droplets of milk, shining and showing a way in the night sky. The supermassive black hole at the center of our Milky Way Galaxy is also called the Sagittarius A. The Milky Way is very old, it has been living for over 13 billion years to be exact.

Here is the image of the Milky Way Galaxy below, and how it is astonishingly beautiful on its full view of the galaxy.

We live in the Milky Way Galaxy, but which way exactly? We live in the Milky Way's Orion Arm, the spiral arm of the galaxy where our solar system is located. The Orion Arm, is named for the Orion constellation, which contains many bright stars and celestial objects found in the arm, such as the Orion Nebula. The Orion Arm also lies between the center of the galaxy which is the Sagittarius Arm and the Perseus Arm which is further from the galaxy.

The Orion Arm has a width of about 3,500 light years, and a length of about 10,000 to 20,000 light years. The solar system is located through the inner edge of the Orion Arm, towards the galactic center.

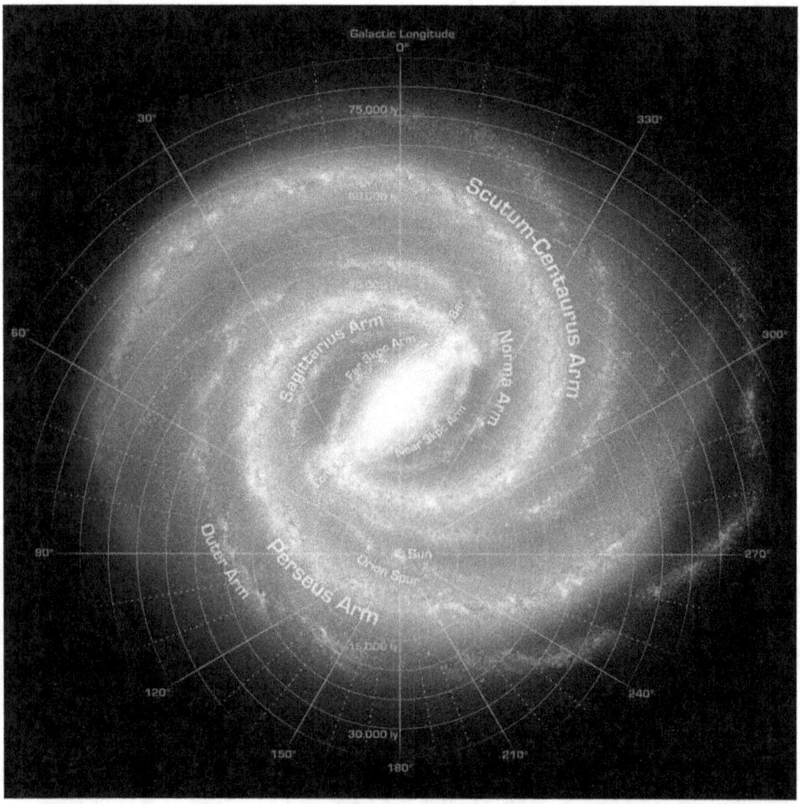

The image you saw on the previous page of the book credited by Wikipedia, it is the Milky Way Galaxy showing the longitude of spiral arms, in each galactic longitude, it contains each spiral arm, and the image shows the Orion Arm, Perseus and other arms.

You may wonder how many spiral arms are also there in the Milky Way Galaxy? Well, there are 12 spiral arms out there in the solar system, also you can also see the solar system in the Orion-Cygnus Arm.

Each longitude degree shows each region and spiral arm in the Milky Way Galaxy shown on the previous page:

- 0°, which is the Sagittarius arm, the spiral arm that is located at the center of our Milky Way Galaxy, the Sagittarius A supermassive black hole is at the galactic center.

- 30°, which is the Aquila Rift, the dark, dusty region that is located between the Scutum-Centaurus Arm and the Sagittarius-Carina Arm, containing clouds of cosmic dust that obscure or hide the light of distant stars.

- 60°, which is the Vulpecula Rift, a dark band of cosmic dust within the Great Rift, the dust lane that divides the Milky Way as seen from Earth, obscuring or hiding the light of distant stars and forming part of the “Dark Zone” or no sign of our stars in the galaxy, just darkness.

- 90°, which is the Cygnus Arm or the Orion Arm, the spiral arm that is located on our solar system, planets orbiting around our Sun, and it is between the Sagittarius Arm and the Perseus Arm.

- 120°, which is the Cassiopeia region, where the Cassiopeia constellation is located within the Perseus Arm, and in the Cassiopeia region, there is also a celestial object out there called Cassiopeia A, it is a supernova remnant, meaning it is an expanding shell of gas and dust left behind from the massive star that has exploded ages ago, and it has the strongest radio source beyond our solar system.

- 150°, which is the Perseus Arm, it is the one of two major spiral arms in the galaxy with the Scutum-Centaurus Arm, which is the second major spiral arm of the galaxy.

- 180°, which is the Auriga, the Auriga is the constellation that is located into the northern celestial hemisphere in the Milky Way, it also contains clusters like M36, M37 and M38.

- 210°, which is the Monoceros Ring, the outer stellar structure of the Milky Way, scientists discovered this structure in 2002. The Ring is also the general region of the galaxy's plane, or imaginary flat surface.

- 240°, which is the Puppis, which is the constellation that is located into the southern hemisphere of the Milky Way and it is the twentieth largest constellation in the universe. The constellation also lies within the plane of the Milky Way.

- 270°, which is the Vela, which is a supercluster hiding within the Milky Way, the Vela supercluster was discovered and identified in 2016, because it was hidden into the Milky Way Galaxy's "Dark Zone" where clouds of interstellar dust obscure the light of distant stars in the galaxy.

- 300°, which is the Crux, or the Southern Cross constellation that is located in the southern sky, at the end of the Milky Way's visible band. The Southern Cross is the most recognizable constellations in Southern Hemisphere countries like South Africa and Botswana, because it has simple four bright stars forming a cross in the night sky.
- 330°, which is the Norma Arm, a minor spiral arm that is made of gas and star-forming regions, which creates an extension near the center of the galaxy and is located between the galaxy's major arms like the Orion-Cygnus Arm. The minor arm is also named after the Norma constellation.

You may wonder what will happen to the future of our Milky Way Galaxy? Actually, there will be some merge happening in the future events of our galaxy. Scientists believe the Milky Way Galaxy and the Andromeda galaxy will be on a collision course and merge into a larger and more supersized galaxy called Milkomeda.

What will this Milkomeda galaxy look like? The Milkomeda will look like an elliptical or oval-shaped galaxy, and the supermassive black holes at the galactic centers of both the Milky Way and the Andromeda will form into a more bigger supermassive black hole which will be at the center of the Milkomeda galaxy. Here is the image of the galaxy collision.

CHAPTER 6.3

PHYSICS & STATES OF MATTER II

THE PROCESS FROM SOLIDIFICATION TO CONDENSATION, AND THE WORLD OF PLASMA

Have you ever saw how the water cycle work in real life? The water cycle is an example of how matter states from water freezing to water evaporating, that's just how matter works in real life. But we haven't learnt about the "fourth state of matter" itself, plasma, and how it relates with the state of matter process.

Remember, on page 99, we explored how gas forms, and on its chapter, focusing on states of matter, we focus on how plasma and other states of matter I haven't mentioned form. This process is also related to the Earth's continuous water cycle, which likely happens mostly on the summer season, when the sun's temperature rises, the water on the earth's ocean, heats up and turns to water vapor, that is the first process of evaporation, and that is how gas forms when I explained the same thing on page 99. But the process doesn't end there, The water vapor in the air forms into clouds, any type of clouds, actually like nimbus clouds, cumulonimbus or stratus. But let's focus on the cumulonimbus clouds, or dark thunderclouds. What are cumulonimbus clouds? Cumulonimbus clouds are different types of clouds that are capable of producing atmospheric electricity inside their cloudy structure, during thunderstorms.

Back to the water cycle process, focusing on clouds turning dark, those are cumulonimbus clouds, all over the sky, obscuring the sun's light, making the place a bit dark. The cumulonimbus clouds then start to rain or release liquid water from the sky, that is the process of condensation (gas to liquid), it is a reversible process of evaporation (liquid to gas).

Or in a different way, cumulonimbus clouds can also start to hail, or releasing ice blocks from the sky, that is the process of deposition (gas to solid), it is also a reversible process of sublimation (solid to gas). The clouds hail, when the water droplets they are producing, raised at the very coldest parts of the atmosphere, causing the droplets to freeze at a lower temperature.

We've learnt how evaporation and condensation process, and now we discover how the fourth state of matter also known as plasma is formed.

Plasma is formed when higher energy such as intense heat or stronger electrical current, is being added to a gas to strip electrons from its atoms, creating to a combination of positively charged ions and negative charged free electrons. Plasma is formed through a process of ionization. The process starts where all matter in the plasma state begin as neutral gas, then energy is being added to the gas, the atoms inside the gas when they are in contact with extreme energy, they start move faster and collide with each other with very strong force

During these atomic collisions, the electrons are being knocked out free from the atoms, becoming free electrons. The atoms that have lost their electrons, are now positively charged ions, meaning they have more protons than electrons that have knocked out by the atomic collision affected by energy. The presence of these charged particles allows plasma to conduct electricity.

Plasma is just a hot, conductive and extremely energized and ionized gas, made of positively charged ions and negatively charged electrons. Plasma also behaves differently because they are strongly influenced by the electromagnetic fields surrounded all over.

They are examples of plasma in our everyday life:

- Stars like our Sun, lightning, the aurora borealis (solar wind), and outer space.
- There is also plasma in our homes, like fluorescent lights (different kind of luminescence that absorbs light or other electromagnetic radiation that emits light at a longer wavelength), flat-screen plasma televisions, and neon signs that are mostly used on tuck shops.

Plasma is also an extreme energized state of matter, that also contains 99% of its state visible around the universe, like stars, the most visible kind of plasma we see in the night sky, while the sun is the biggest kind of plasma in the day. Plasma is not just an ionized gas that has extreme amounts of superheated energy such as high voltage electricity that knocks electrons off atoms inside the gas. Plasma is also blood in our body, Mind-boggling, isn't it?

Is this fourth state of matter really blood? But this is a different kind of plasma. Yes, some other sources agree with the question. Let's discover how plasma interacts with our body's circulatory system as well. Plasma is a yellowish liquid inside our blood system, which makes up a percentage of 55% of the whole blood in our body, amazing. The whole blood in our body is made up of a combination of plasma and other blood cells like erythrocytes (red blood cells), leukocytes (white blood cells), and platelets. So, actually, no, the plasma that is the fourth state of matter is not blood, there is a difference between the two, but they share the same name from the Greek word "plassein" which means forming or molding. Let's move from the biological plasma back to the physics plasma. This time, we focus on the history of plasma, who discovered plasma, and how did he discover this 4th state of matter?

We travel back in time, to 1879, where British physicist Sir William Crookes observed a glowing, electrically conductive gas inside his Crookes tube and he called it "radiant matter" because of its extreme luminosity and brightness. Then, 18 years later to 1897, British physicist J. J. Thomson, the mastermind behind the plum pudding model, identified the electrons' ionized nature.

Then 31 years later, to 1928, American physicist Irving Langmuir first coined the term or came up with the word "plasma", reminding of blood plasma, carrying blood cells, because ionized gas consists of electrons and ions. What is the importance of plasma in our universe? Plasma might be hazardous, an example of its hazard is a lightning strike. Plasma also plays an important role as it is themost common state of matter that is involved in the whole universe, like stars, the solar wind, or outer space.

Here are the images of examples showing different kinds of plasma: lightning, the plasma globe, the Sun, and space.

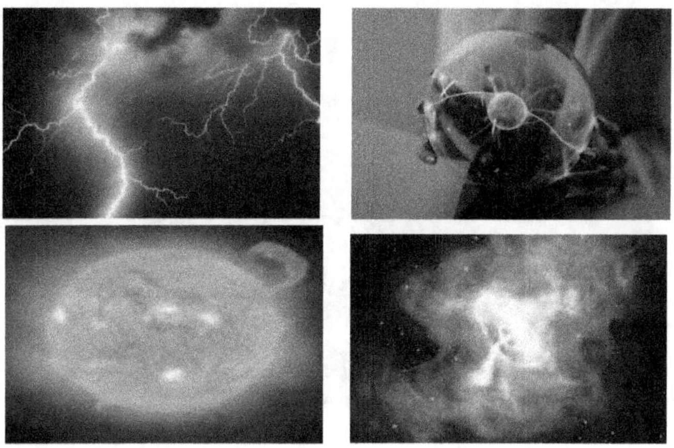

CHAPTER 7

EVERYTHING YOU NEED TO KNOW ABOUT PHYSICS & MORE METRICS

Have you ever wondered what is the system that helps you measure different sizes of objects and different volume of liquids? Obviously, it is the metric system. The metric system or the International System of Units, is a decimal-based system that measures different quantities. In this chapter, we will discover more about the metric system, and units I haven't explained before on page 32, the chapter focusing on the rivalry between the metric and the imperial system in this book. The metric system is the victorious decimal-based system that is mostly used to measure different quantities, like mass, length and time.

The reason behind its victory, because the metric system of units is used internationally, it was first discovered in the country of France, during the French Revolution, in 1790. Let's discover more metric units and derived units, and how the metric system shapes the world more than the American imperial system.

CHAPTER 7.1

PHYSICS & RATES OF CHANGE

DERIVED UNITS AND DIFFERENT RATES THAT CHANGE OVER TIME

Have you ever wondered how rate changes over time, metaphorically? In the world of mathematics and physics, rate is a characteristic of a derived unit. What are derived units? A derived unit are a different type of unit that are formed by a combination of SI base units (e. g., meter) through mathematical operations like multiplication or division.

Derived units are used to measure physical quantities that are not fundamental, like area, volume, density, pressure and velocity. These units are also called "rates of change" units in mathematics. These derived units also a play an important role of helping mathematicians and physicists to solve formulas using these derived units like force it is measured in Newtons but it has a combination of base units between mass and acceleration through multiplication.

These are the different types of derived units in the metric system of units:

- Speed is a derived unit that is formed by dividing a distance unit (meter) by a time unit (second), which is meter per second or m/s, showing how far a car or an airplane travels in a certain amount of time.

- Area is a derived unit that is formed by multiplying the length unit to the exponential power of two (meter × meter = m^2).
- Force is a derived unit that is formed by multiplying the mass unit (kilograms) by acceleration unit (meters per second), in mathematical form, it is $kg \cdot m/s^2$.
- Volume is also a derived unit that is formed by multiplying the length unit to the exponential power of three (meter × meter × meter = m^3).
- Pressure is a derived unit that is formed by dividing the force unit (newton) and the area unit (square meter), or in mathematical form as N/m^2
- Electric power (measured in watts) is a derived unit that is formed by dividing the energy unit (joule) by the time unit (second) or in mathematical form as J/s.
- Frequency (measured in hertz) is a derived unit that is formed by multiplying a time unit to the power of negative one.

You may wonder what is the use of these derived quantities? Yes, the use of these derived quantities is to solve difficult mathematical problems. Here's how physicists and mathematicians use the power of derived units to find a solution on their formulas.

For example: Physicists calculate the volume of a liquid

V = l x b x h (length × breadth × height)

= 10 × 10 × 10 cm^3= 1000 cm^3 which is also equivalent to 1000 *ml or 1*

Discovering more about derived units will help you become a maths and a physics genius one day, these derived units from the SI are the easiest but hardest tools you can use to solve problems in both mathematics and physics, today. You can not just use derived units to solve problems, but even formulas and mathematical equations you can use as well.

There are also different units that change over time. Time plays an important role in speed and velocity.

- Acceleration (changes in speed), measured in meters per second squared (m/s^2)
- Velocity, is the measurement of speed in a direction of motion, it is a vector quantity, meaning that it has magnitude and direction. $V = d/t$

You may wonder what is the difference between a scalar quantity and a vector quantity? A scalar quantity is also a different type of quantity that can be described as a real number accompanied by a unit of measurement for example; 1 km which in a word form refers to one kilometer. When scalars have both magnitude and direction, that is a vector. Example of a vector refers to 1 km north (north which indicates a type of direction).

So here is the clear difference between a scalar quantity and a vector quantity

- A scalar quantity has the size of a quantity (magnitude) but it has no direction. (example: 10 km)
- A vector quantity has both the magnitude and the direction (example: 10 km north or 10 km east)

Velocity is actually the vector quantity of speed, why is it the vector quantity of speed? Some other people think velocity and speed are the same thing, it is actually wrong. There is a difference between velocity and speed. Speed is a derived unit that is measured distance over time and a scalar quantity, because it has no direction. But velocity on the other hand, when the speed of the objects stays constant, the direction changes, so there is clearly a direction in velocity where the speed stays the same, but the direction changing, that is why velocity is a vector because it has both the constant speed (magnitude) and the changing of direction.

If you still don't understand, here's how we solve the problems of calculating both the speed equation and the velocity equation (which is the same equation), but different quantities, and the clearly see the difference between the scalar and the vector quantity. Calculating the speed equation (scalar quantity, a type of quantity has the magnitude accompanied by a unit of measurement, but has no direction) The problem: If a car travels at a distance of 20 km, in a duration of four hours, what is the average speed?

$S = d / t$

= *the distance is 20 km divide the distance by 4 hours*

= *the average speed is 5 km/h (example of how we calculate a derived unit of speed which is a rate of change over time)*

5 km/h with no direction, is not a vector quantity, but the equation that I calculated proves that speed is a scalar quantity, not a vector quantity. Calculating the velocity equation (vector quantity, a type of quantity has both the magnitude and direction accompanied by a unit of measurement) The problem: If a train travels at a distance of 40 km, in a duration of ten hours, what will be the average speed?

V = d / t

= the distance is 40 km divided by 10 hours

= the average speed is 4 km/h south (example of a vector quantity, adding the direction the train travels).

We have learnt how the derived units and its types of quantities work on the metric system, and now we move to the imperial system. The imperial system is the type of non-decimal system of measurement that is mostly used in the United States of America, Liberia and Myanmar. The imperial system has different units of measurement that are similar to the metric system. Here is a comparison between the two systems.

Comparison between the metric and the imperial systems

- The metric system has seven fundamental metric SI base units like mass (kilograms), length (meters), time (second), temperature (kelvin), electric current (ampere), amount of a molecular substance (mole), luminous intensity (candela).
- The imperial system doesn't have base units, but has similar units from the metric system, like mass (pounds), length (inches), time (second), temperature (fahrenheit), electric current (ampere), amount of a molecular substance (mole), luminous intensity (candela).

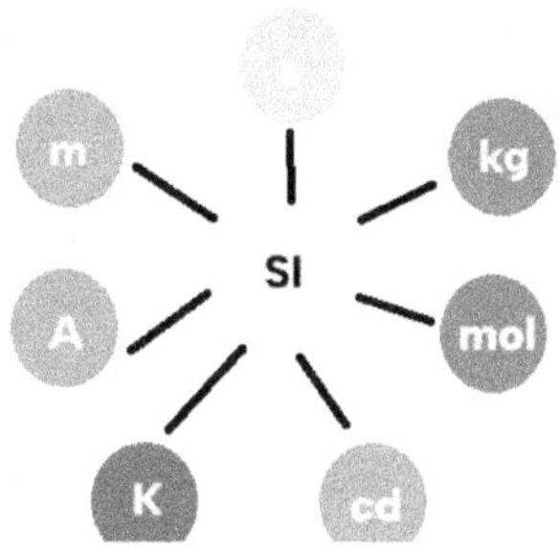

The derived units focusing on the imperial system, uses similar quantities I have explained from the metric system's derived units. Those quantities, are speed (miles per hour), area (square foot and square miles), volume (cubic foot), pressure (pounds per square inch). The metric system and these derived units that "change" over time have revolutionized the most parts of the world, except the United States, Liberia and Myanmar.

Here is a scientific fact, about the history of our metric system we used today. In 1791, there were two French astronomers, Jean-Baptiste Delambre and Pierre Méchain travelled to measure the distance between the North Pole and the equator, to create the first unit on the metric system known as meter, which is one ten-millionth of the distance between the North Pole and the equator.

CHAPTER 7.2

PHYSICS & TEMPERATURE

THE CELSIUS, KELVIN, FAHRENHEIT, RANKINE SCALE, AND THE FORMATION OF HEAT

Have you ever wondered why there is cold and why there is hot in our world, and have you also wondered why is it important in our everyday lives as well? Temperature is everywhere, from hot electrical devices, to cold refrigerators. We, humans also have our average body temperature at 37° C. It can be converted to different temperature units like fahrenheit and kelvin.

Calculating of finding a solution of converting from celsius to both fahrenheit and kelvin.

Starting with the Celsius to Fahrenheit formula

$°F = (9/5 \times °C) + 32$

$= (9/5 \times 37° C) + 32$

$= 66.6 + 32$

$= 98.6°$ *F (average body temperature in Fahrenheit scale)*

After, the inverse formula of the previous formula, which is also the Fahrenheit to Celsius formula

$°C = (°F - 32) \times 5/9$

$= (98.6 - 32) \times 5/9$

$= (98.6 - 32) \times 0.56$

= 66.4 × 0.56 = 37° C (average body temperature in Celsius scale)

Now, to the Celsius to Kelvin conversion formula

°K = °C + 273.15 (absolute zero temperature, it is -273.15 deg Celsius, not positive)

= 37 + 273.15

= 303.15 °K (average body temperature in Kelvin scale)

You may also wonder what is that 9/5 for on the Celsius conversion formulas? Well, in Celsius scale, water freezes at 0° C, and boils at 100° C, which is at a range of 100 degrees. In Fahrenheit scale, water freezes at 32° F, and boils at 212° F, which is a range of 180 degrees. The fraction is 180/100, the fraction in its simplest form is 9/5. Well, the meaning is that 180 divided by 20 is 9, and 100 divided by 20 is 5. The 9/5 in the formulas simply means a 1 degree change in the Fahrenheit scale.

The number 20 is the HCF (Highest Common Factor) because it is the largest number that divides two or more numbers leaving without a remainder.

So, let's go back to the topic's question: why is temperature important to our daily lives? Some people how can temperature be important, it is just changes from hot to cold, from hot to cold and so on, like it does on seasons throughout the year, autumn and spring, normal temperatures, but summer with hotter days, with more rain expected and longer days, and winter is very cold and dry with shorter days. Some people don't perceive something important about temperature in seasonal changes.

Before we find the answer to our question, let's explore more about the history of these temperature scale units like Kelvin,

Celsius and Fahrenheit. Let's travel back in time to the timeline of how the Celsius scale was invented by the scientist behind the name, Anders Celsius.

We go back in time to the year 1742, where Swedish astronomer Anders Celsius, conducts an experiment by creating a temperature scale at his lab, where the freezing point of water and the boiling point of water were at a normal atmospheric pressure.

Celsius initially found out in his temperature scale that the boiling point of water starts at 0 and ends with the freezing point of water ends at 100. But in the later years, scientists reversed the scale, as 0 is now the freezing point of water and the boiling point of water at 100. The scale around the world is also called the centigrade scale because it has one hundred (100) Celsius degrees, where centi- means 100 and gradus means steps in Latin.

Here is the Celsius-Fahrenheit-Kelvin-Rankine scale diagram showing data from absolute zero to the hottest temperature possible, similar to the diagram shown on page 41, the difference is that it shows a different temperature scale called the Rankine scale, which is the scale you will discover more about later on.

What is the Kelvin scale, and how was it invented, and who invented this scale? The Kelvin scale was invented by the scientist behind the name, Lord Baron Kelvin. How did he invent the Kelvin scale?

In the year 1848, Lord Baron Kelvin, invented the Kelvin scale, by proving or theorizing an absolute temperature scale based on the connection between the gas' volume and temperature, meaning that the gas volume would become zero at a Celsius temperature of -273.15° C, which he called absolute zero in a Kelvin temperature of 0 K.

The Kelvin scale uses the same degree as the Celsius intervals, but starts with the exact zero point, representing the absence or nothingness of heat energy and molecular motion, making it an absolute scale with no negative numbers. The freezing point of water is at 273 K, and the boiling point is at 373 K.

The Kelvin scale uses the same degree as the Celsius intervals, meaning that a change of 1 degree Celsius is equivalent to a change of 1 degree Kelvin. The Kelvin scale is an absolute temperature scale that begins with the most coldest temperature possible which is absolute zero at 0 K, and in Celsius interval it is -273.15° C, most coldest ever in the universe. What is absolute zero, exactly? Absolute zero is most lowest and coldest temperature ever possible at a lower temperature of -273.15° C, 0 K, -460° F and 0 R (same interval with Kelvin).

Absolute zero on earth cannot be reached. Why? In the third law of thermodynamics, absolute zero cannot be reached in a limited number of steps or limited number of energy. In theory, absolute zero can be reached in infinite energy, where it causes to stop molecular motion and heat.

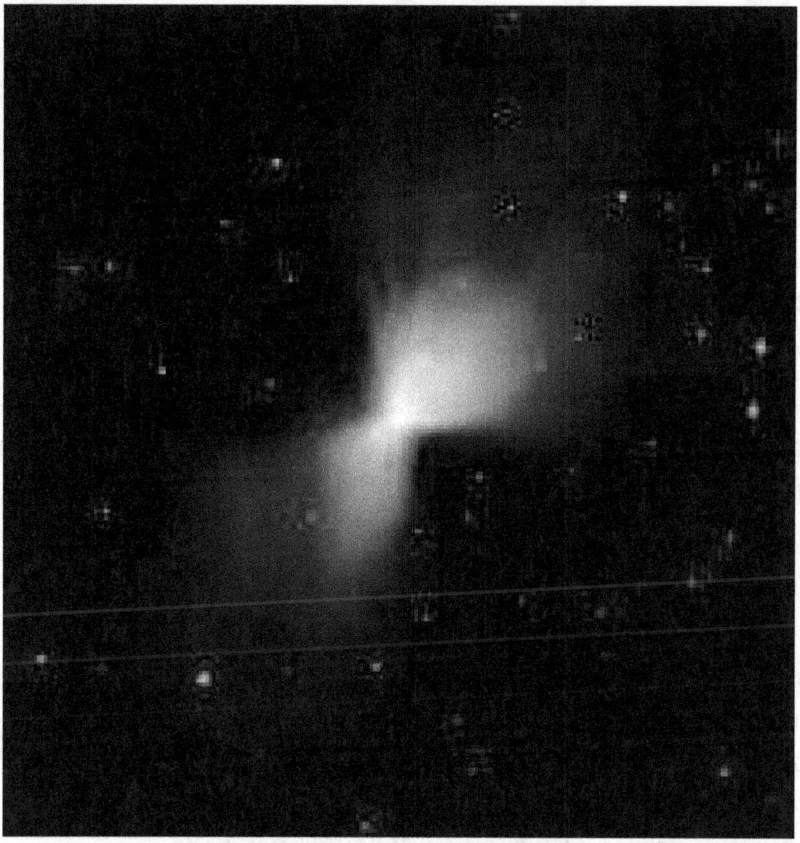

This image here shows the Boomerang Nebula was discovered to be releasing gases in outer space, at a speed of half a million or 500,000 kilometers per hour (km/h), in the last one and a half thousand years. The average temperature of the whole universe today is 2.73 K or -270° C, or -454° F. Scientists believe that the continuous expansion of the universe in the future, its temperature is cooling down over time. But there is no way, the most coldest temperature possible, absolute zero reach, even on cooling devices we have on our homes, like refrigerators, or ice coolers. Next, we journey to the invention of the Fahrenheit scale, and how it was developed, by the scientist who also behind the name, Daniel Gabriel Fahrenheit.

In the year 1724, German physicist Daniel Gabriel Fahrenheit, invented the temperature scale that its freezing point of water begins at 32° F, and its boiling point of water is at 212° F, at normal atmospheric pressure, which is the Fahrenheit scale.

How was the Fahrenheit scale invented? Before the creation of the scale, Fahrenheit was researching more about Ole Roemer's temperature scale, a different scale that has a freezing point of water at 7.5 deg and a normal body temperature of 22.5 deg.

He was fascinated by Roemer's scale, and adapted his temperature scale, and made changes by avoiding fractions from Roemer's scale, by starting originally from the zero point which is the temperature of a solution of ice, water and ammonium chloride

Then, Fahrenheit multiplied all the numbers on his scale by four, and the multiplication later created the most known points of the Fahrenheit scale today, the freezing point at 32° F, to the boiling point of 212° F. While the body temperature in the Fahrenheit scale, is about 98.6° F, the same number that I have found when converting from Celsius to Fahrenheit on page 185.

The Fahrenheit scale is also a temperature scale part of the imperial system of measurements mostly used in countries where they use the imperial system such as United States of America, Liberia and Myanmar. The reason why the United States mostly use the Fahrenheit scale, the obvious answer is because it is part of the imperial system, but it is more than that. The story started in the 20th century, when scientists Halsey and Dale chose the Fahrenheit scale more than the Celsius scale, because of its range of 0 and 100 degrees, like when the thermometer shows 0° F means that it is a frosty and very cold winter day, and during summer when it shows 100° F, means that it is a hot summer day.

100° F is not the boiling point of water, but 4 degrees forward after our body temperature which is 98.6° F.

You have explored how both the Kelvin, Celsius and Fahrenheit scales were invented by three scientists, Anders Celsius, William Thomson, 1st Baron Kelvin and Daniel Gabriel Fahrenheit.

Now, we discover the temperature scale that you haven't heard of before, which is the Rankine scale. How was this Rankine scale invented? The Rankine scale is a different type of absolute scale of thermodynamic temperature, meaning it is a scale similar to a Kelvin scale, starting from absolute zero which is the exact zero point of the scale.

The invention was first proposed in the year 1859, where University of Glasgow physicist W. J. M. Rankine, who named his absolute temperature scale with his name. The Rankine scale's one degree difference is also equivalent to the Fahrenheit scale's one degree difference like the Celsius and the Kelvin's one degree difference equivalence. The Rankine scale has a freezing point of water that has a temperature of 491,67° R, and a boiling point of water that has a temperature of 671.65° R. The conversion formula from the Celsius scale to the Rankine scale is ° R = (°C × 9/5) + 491,67.

Let's calculate the average body temperature, this time in a different conversion formula from Celsius scale to Rankine scale

° R = (°C × 9/5) + 491,67

= (37 × 1.8) + 491,67

= 558,27 °R (average body temperature in Rankine scale)

So, let's find the answer to that question: Why is temperature important in our daily lives on earth? Some people say it is just cold and hot, how can it be important. Let's discover why is temperature important.

Temperature is important because it is one of the most influential force of nature in the world, affecting health, agricultural productions and plant growth.

The actual importance of temperature that it is involved during the water cycle process, it means that the sun produces a hot temperature onto the earth's oceans, causing the water in the oceans to heat up at a high temperature to evaporate, then water vapour in the air creates clouds in the air, then precipitation takes place, and fills the place with water, and temperature is also involved in the photosynthesis process, helping plants to produce food and nutrients. Temperature is also crucial on the Earth's biosphere.

Another question you may wonder, what is the biosphere, and how does it connect with temperature? Well, the biosphere is the ecosystem that has a collection of living organisms, like flora (plants) and fauna (animals), even us, humans. Temperature plays an important role in the biosphere, because most life and us humans require liquid water, for survival. It is also the reason behind the water cycle, and photosynthesis. Other living organisms cannot survive at extreme temperatures of below 0° C, and above 40° C.

The greenhouse effect, is a process where greenhouse gases in the atmosphere, such as water vapor and carbon dioxide (CO_2) traps the heat from the Sun, and returns back to the Earth's surface. So what will happen if there was no existence of temperature on earth? Well, what will happen is that the Universe will also cease to exist, because the universe was created by huge amounts of heat and strong radiation during the events of the Big Bang.

Temperature is actually a measure of kinetic energy of subatomic particles within a system. When temperatures increase, particles move faster and vibrating, producing higher amounts of energy, causing the substance more hotter. And when temperatures decrease, particles move slower, causing the substance to be colder.

And if there was no temperature on earth, all life around the planet will be extinct for eternity, no respiration, no photosynthesis, no Sun as well, and no Great Britain, no Australia and no South Africa, and no continents and countries, and no everything on earth actually.

And no chemical reactions as well, chemical reactions are the possible ways to produce the energy of particles. Without temperature, particles would not able to produce kinetic energy, and all particles will remain still, lack of motion, just stillness.

There would be obviously no water cycle on our Earth, causing the planet to be dry, and filled with nothing but dust and rocks. Even the existence of temperature is important in our daily lives, thanks to God, temperature exists, if it hadn't been existed, there would be no humanity right now.

The universe couldn' t have been created as well, because of the non-existence of temperature, it would have been an empty void of nothingness with no active chemicals to form a Big Bang. With temperature, the formation of planets and stars, involves immense and powerful energy, with particles at different temperatures. But without temperature, there would be no existence of these celestial matter such as stars, lack of energy. It would need particle energy to form stars and planets. That is why temperature's existence is the most important part in our daily lives, and all of life out there on earth. Here is another question: how does heat form, and what is also the importance of the heat?

Heat is a different type of energy that forms from the movement and vibration of particles in matter. It is also a process that can be formed through friction, burning, and violent chemical reactions, with higher amounts of heat.

When particles move faster and vibrate more extremely, they build huge amounts of heat, and when particles move slower, and vibrate slowly, they decrease heat, and increase lower temperatures.

- Heat is formed when the kinetic energy of the particles are being produced, the faster the particles move or vibrate, the greater the kinetic energy, and also the hotter the substance.

- As these rapidly moving particles collide with slower moving particles, they transfer the heat to the slower particles, causing them to build more energy, and the substance even more hotter

- Friction is caused by heat, because rubbing objects together rapidly, causes the objects' particles to move faster, building heat.

- Chemical reactions occur when the burning or combusting release energy in the form of a thermo (heat) and photo (light).

What is more to heat in our world, and what does it actually do atomically? What is heat transfer? Heat transfer is a process where particles move rapidly with thermal energy from a higher temperature part of the object to the lower temperature part of the object. How does heat transfer actually work?

Heat transfer works through four main processes; conduction, convection, radiation and advection.

- Conduction: the heat transfer process where heat is being transferred through direct contact. Example:
- Convection: the heat transfer process where heat is being transferred through the movement of fluids like liquids or gases. Example: The water from the pot boils, releasing gas.
- Radiation: the heat transfer process where heat is being transferred through invisible electromagnetic waves. Example: The microwave reheats the food.
- Advection: the heat transfer process where heat is being transferred through the movement of fluids from one location of another. Example: The smoke from home is being released into the air.

Physics and temperature have shaped our lives once again, it has revolutionized life and its unknown future. Some say temperature is just hot and cold but they also do not know that temperature is really important in life. Temperature has been with us since the Big Bang. But we also discovered the comparison between the existence and the non-existence of temperature, and we found out that without temperature, there would be no Big Bang, no life and no Earth, and obviously no particles or no anything in fact, just a void of nothingness. Physics has shaped our lives once again with its important connection to the imaginative world of hots and colds, or temperature as we know it.

CHAPTER 7.3

PHYSICS & MASS

WEIGHTS AND MEASURES, METRES, KILOS, CENTIS AND MORE

Does size really matter? Why do other people ask this rhetorical question? This question has a lot of answers, some say size doesn't matter, others say size matters. We discover into the world of weights and measures, and find out why mass is in our universe. What is the importance of mass and its properties? Let's find out how and why.

Mass has been in our world since the Big Bang. What is mass, exactly? Mass is the measure of quantity that matter has. Does mass create itself?

Yes, but mass is made out of matter (substance that has mass and occupies space or volume). Who discovered mass and why is it in our universe since the Big Bang? Energy is actually the reason why there is mass in our universe, containing leptons and quarks. *Discover what leptons and quarks are on pages 71 and 72.*

A lepton is an elementary particle that has a ½ spin similar to other subatomic particles like electrons. A quark is a fundamental particle that creates electrons and neutrons from electric charges. Mass is everywhere, how much does the Earth weigh with so many people on its surface, how?

Even at the court, during a case, they like to use the Lady Justice balance scale, which is a symbol of mass in physics, but in law, it is a symbol of balancing humanity and actually weighing evidence collected by lawyers and other law specialists. Let's step out from law, this is not about law, this is about mass-based physics. But the Lady Justice's balance scale is a representation of mass, weighing and measuring. Why is mass so strong and powerful depending on the planets' sizes, why?

People wonder what are weights and measures, and how do they work, and why is it crucial and involved in our everyday lives? Let's initiate with weight. What is weight? Weight is the measure of an object's mass or heavy size, which is used in physics and everyday mathematics, e. g. imperial units like pounds, tons and metric units like kilograms.

Measures on the other hand, it is a way to determine or observe the size, dimensions of some objects, e. g. physicists and mathematicians use metric units that measure light objects but not heavy ones, like meters for length, kilometers for distance, or liters for volume of water. Other people ask this question: Why is it important, and why is it involved in our everyday life?

Weights and measures are crucial in our everyday life, because people have to observe their mass, how many pounds they have gained or how many pounds they have lost, and weight-based tools like body weight scales, these tools are helpful after exercise to see how many pounds you have lost from exercise, or before exercise, how many pounds you have gained.

Weights and measures are also crucial in everyday scientific research to scientific work environments, when looking for some important piece of information or data. Why is it also involved in our everyday lives? Weights and measures are also involved in our everyday lives, because of fair trade in the economy, ensuring that the goods are in the correct amount of units.

But what is more to mass outside our home planet? Well, actually, the Sun takes over the mass quantity of the whole solar system. What I mean is that, the Sun has 99.85% mass of almost the whole solar system, and it is over 300,000 times massive than the Earth's size.

Astronomers once discovered Quipu, a massive cosmic structure that is 200 quadrillion (2 × 10^15) times bigger than the Sun, it is found in 63 galaxy clusters. Maybe, Quipu is the massive and heaviest cosmic object that is in the universe, maybe.

Okay, maybe not, because Quipu is not the biggest thing here in our universe, it is actually the Hercules-Corona Borealis Great Wall. The Hercules-Corona Borealis Great Wall is actually the largest known structure in the universe, and it is 10 billion light years across our solar system, and it is actually the biggest galactic filament in our universe.

Let's focus on the most easiest and important thing in the metric system: the metric prefixes like kilos, centis, millis, teras, gigas and more.

But why are these prefixes part of the metric system? These metric system prefixes are also related to the mathematical exponential system, and the reason behind the exponential connection with metric prefixes, because the International System of Units (SI) or the metric system is a decimal based system, and the prefix is measured to the power of 10.

Another reason behind the exponential connection with the metric prefixes, the metric system itself uses a base-10 system, it means that the units in the metric system converts using a simple multiplication or division by the power of 10. Students at primary and high schools learn the metric system and practice to understand how the base-10 conversion work.

List of the metric prefixes in the International System of Units: *Discover other metric prefixes on pages 37 and 38.*

- Small multiples of the metric prefixes
- Deci (10^-1), 1 ÷ 10, e. g. decimeter (0,1 m, tenth of a meter, or tenth of any base unit)
- Centi (10^-2), 1 ÷ 100, e. g. centimeter (0,01 m, hundredth of a meter)
- Milli (10^-3), 1 ÷ 1000, e. g. millimeter (0,001 m, thousandth of a meter)
- Micro (µ, 10^-6), 1 ÷ 1000000, e. g. micrometer (0,000001 m, millionth of a meter)
- Nano (10^-9), 1 ÷ 1000000000, e. g. nanometer (0,000000001 m, billionth of a meter)
- Pico (10^-12), 1 ÷ 1000000000000, e. g. picosecond (0,00000000001 m, trillionth of a meter)
- Femto (10^-15), 1 ÷ 1,000,000,000,000,000 , e. g. femtosecond (0,000000000000001 m, quadrillionth of a meter)
- Atto (10^-18), 1 ÷ 1,000,000,000,000,000,000 , e. g. attosecond (0,000,000,000,000,000,001 m, quintillionth of a meter)
- Zepto (10^-21), 1 ÷ 1,000,000,000,000,000,000,000 , e. g. zeptosecond (0,000,000,000,000,000,000,001 m, sextillionth of a meter)

- Yocto (10^-24), 1 ÷ 1,000,000,000,000,000,000,000,000 , e. g. yoctosecond (0,000,000,000,000,000,000,000,0001 m, septillionth of a meter), and so on

These metric prefixes change the quantity of these metric units, these small multiples use the division operation to be divided by the power of 10. Physics and mass have revolutionized the world of weights, measures and how metric prefixes work, again. But as we conclude this sub-chapter, and head to another sub-chapter, to learn something new and something more about metrics and its inner workings like energy, which is next.

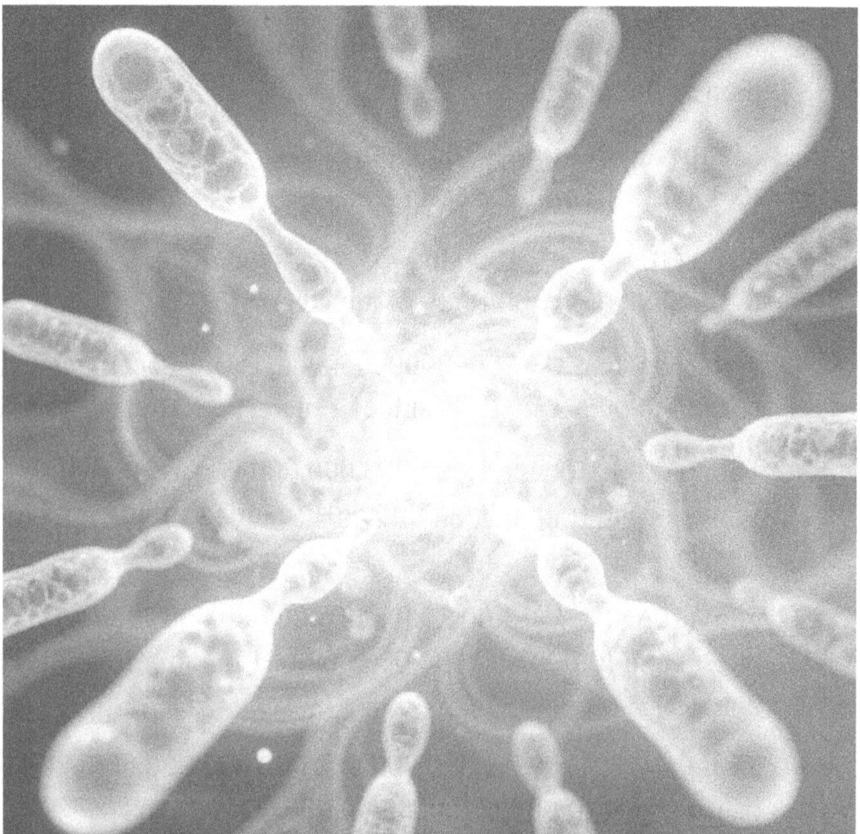

CHAPTER 7.4

PHYSICS & ENERGY

JOULES, KILOJOULES AND CALORIES FROM FOODS AND THE PHYSICS BEHIND IT

Have you ever wondered what is energy, why is it involved in our world, and what would've happened if energy never existed? This is an interesting question, to most people. Energy is the ability to work, like exercising, burning calories from your everyday diets.

Another question, is who discovered energy? This is the sub-chapter, where you will find all the answers of why energy is in our world, and why is it crucial in our everyday life.

The word "energia" was termed by Ancient Greek physicist Aristotle in 4th century BC, which meant activity or work. The English translated word of energia which is used today as the word energy was termed by American physicist Thomas Young, the master behind Young's modulus, which means it is a type of modulus in physics that is a measure of elastic energy, or stretchy force.

Why is energy involved in our world, I meant in our universe, but why? Energy is actually involved, because appliances and electrical devices we use are powered by different kinds of energy.

Like heat energy, which produces large quantities of heat (example: a stove, it produces heat to boil the water in the pot, or to cook the chicken), light energy (example: a light bulb, produces both heat and light energy forming into one type of energy, which is electrical energy).

Energy is measured in these metric units called joules (J). Do you actually know what joules are? Joules are these metric units you see on your snack ingredients at the back of your packet (example: Carbohydrates – 2.6 kj or 2600 joules at the back of your packet or your Corn Flakes carton box or any product you buy from your supermarket).

Why is energy crucial in our everyday lives as well? Energy is crucial because it is everywhere, energy is in us, energy is in every electronic device, and every heaters during the cold winter season, making sure our homes are in a comfortable state.

And yes, most importantly, energy helps us to protect our communities from danger, like security cameras powered by electrical energy, or even energy itself can be dangerous like fires to electrical shock hazards, from lightning to radiation as well. What do I mean by energy is in us, well, we, humans we eat food, we drink water, in order to re-energize ourselves from tiredness, exhaustion and fatigue as well, and even exercising to build energy and burning calories and even kilocalories, even sleep helps us to re-energize, building more energy.

In our human body, we have chemical energy, like in our stomachs, full of stomach acid or hydrochloric acid which builds when we eat, and digests the food we have eaten, and does the job, causing excreting and does things like that.

Energy in our everyday lives, helps us to build mental strength, stamina and even helps to stay healthily, happy and comfortable, using electronic devices like televisions producing both sound, heat and light energy to watch your favorite shows and more that energy has in store in our everyday lives.

Yes, energy is measured in metric units called joules, named after the British physicist James Prescott Joule, who did his work of finding out the fundamental works of energy and how it actually works and why is in our world in the 16th century. Energy is also termed in Einstein's mass-energy equivalence, which means that German physicist Albert Einstein proved that energy is equivalent to mass multiplied by the speed of light squared or in equation form which is known around the world today as "$e=mc^2$".

But this time, we are focusing on the units that measure the energy in our foods, and energy in us, joules, kilojoules (equivalent to 1000 joules), and calories (how much energy wereceive from foods). A joule is a unit of energy by the International System of Units or the SI, which is equivalent to a force of a newton moves a distance of a meter divided by a unit of time squared and in the direction of the force, here is an equation: $J = kg \cdot m^2 / s^2$.

What are calories exactly? Calories are a unit of energy that measure the energy in the food, and the energy yout body functions like breathing, thinking and moving, and improves your bodily functions. Calories provides fuel for your body, which is being contained in carbohydrates, fats, proteins which are three macronutrients (large nutrients in your body).

What is the physics behind energy in our everyday life anyway? A calorie is actually equivalent to 4.184 kilojoules, or 4,184 joules, and that's not all. You may wonder, that did British physicist James Prescott Joule discover the calorie unit?

No, exactly, another chemist from France, named Nicolas Clément-Desormes discovered the calorie unit in the 1820s, he proved that a calorie is an amount of energy that raises a temperature of 1 kilogram of water by 1 degree Celsius, and also found in our human bodies, when we consume food, and releases different units of it inside our bodies, and helps improving with the bodily functions.

While James Prescott Joule in the mid-1500s, proved that heat is a form of energy that is being compared between mechanical work or energy and mechanical heat, which he termed as the concept of the mechanical equivalence of heat and energy. In this world full of energy, we have many kinds of energy, here is a list below:

- Biological energy: a type of energy produced by humans and animals, from eating food, and exercising, or running for animals.
- Kinetic energy: a type of energy that produces movement, like a rocket flying towards the sky, or a car moving, important type of energy required on Newton's laws of motion.
- Potential energy: a type of energy that is stored for the object to move in future, like a kid who is riding a toy motorcycle, who uses his legs to produce energy, and then cause the motorcycle itself to move in the future, when he releases energy.

- Elastic energy: a type of energy that is stored by elasticity, when an elastic rubber band is expanding, it builds energy, and when it contracts, it falls back into its original shape.
- Heat energy: a type of energy that produces heat, when friction between the particles or objects increases, it produces heat.
- Electrical energy: a type of energy that produces electricity, using negative particles known as electrons that releases huge amounts of heat and light to produce electricity, but it is hazardous.
- Renewable energy: a type of energy that is being renewed and it is irreplaceable when producing electrical energy, example: solar energy or photovoltaic energy, energy from the Sun, water-powered energy or hydroelectric energy, energy from water, and finally, wind energy.
- Non-renewable energy: a type of energy that is replaceable, and not being renewed, not from nature. It is a type of energy that is harmful to the environment, for example; coal that is being produced on factories to release harmful fumes, that harm the environment.

So, energy has powered our minds, and sparked imagination energy, to imagine more about the works of physics, and its surroundings in our curious world. Discovering what energy actually is, why is it involved in our world, and why is it mostly important? Physics and energy have an empowering connection that would enlighten your curiosity with light energy, if you know what I mean. But, seriously, energy does build and improve our health in our human bodies, and comforts our home, and brings us joy with television all from the help of energy. Thank you, energy.

CHAPTER 8

EVERYTHING YOU NEED TO KNOW ABOUT PHYSICS & THEORIES

Have you ever wondered what created the universe, or what does $e = mc^2$ mean, exactly? It all comes down to theories, the Big Bang theory, the theory of everything, and the theory of relativity which was explained by many legendary physicists. You may wonder, what is a theory, exactly?

A theory is a set of principles that scientists explain some facts, and those principles can be used to make predictions, and developing important evidence. But, there is a question: are these theories proven by scientists true or not?

Discover these theories, and how legendary scientists like Albert Einstein, the master behind the theory of relativity, and others who explained the hypothetical theory, meaning the scientists explained if the theory is true or not, in this upcoming sub-chapter on the next page.

CHAPTER 8.1

PHYSICS & THE GENERAL THEORY OF RELATIVITY

IS GRAVITY NOT A FORCE, BUT REALLY THE FOUR-DIMENSIONAL SPACETIME?

Have you actually wondered what Einstein's general theory of relativity is? Einstein's general theory of relativity, is a theory that was proven by Einstein, that gravity is not an invisible force that pulls objects towards each other, but a four-dimensional spacetime caused by mass and energy.

But what does it actually mean? So, gravity is not a force, but a spacetime curvature that holds planets and other celestial objects, to look like they are orbiting. This theory is connected through cosmic events that happen around the universe, like planets orbiting around stars, gravitational lensing from black holes. Massive objects like massive stars warp the fabric of spacetime, like objects follow the curves, meaning they orbit around the massive object, which is observed as gravitational force.

Here is the picture below of the spacetime curvature as the fourth dimension, explained by Einstein in his theory, which might look science fiction, but it is actually reality, sort of.

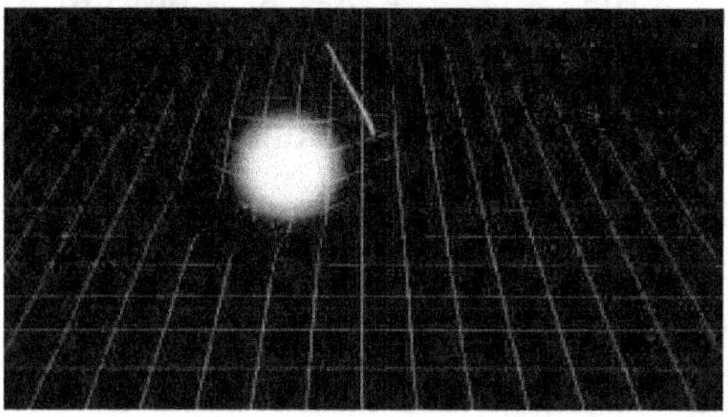

In Einstein's general theory of relativity, as you saw the image on the previous page, it is shown that massive objects like that one huge celestial object you saw, acts as the core of the curvature, and warps the fabric of spacetime, and smaller objects or planets orbit around that massive object, it is similar to the Earth orbit around the Sun, and ends its revolution for a year or 365 ¼ days, and the Moon orbiting around the Earth, and ends its revolution every month. But how did Einstein theorize his general relativity?

Einstein actually theorized his general theory, in 1907, on his thought experiment where he found out that gravity (the force that pulls objects towards each other) and acceleration (the change in speed) are the same, and obviously identical. That's when Einstein made an idea of a fabric of spacetime, that is warped by both mass and energy, and he used Riemannian geometry, a different kind of geometry that measures the curves within a geometrical shape with the assistance of mathematician, Marcel Grossmann to prove how the spacetime curvature influences the movement of objects, showing this theory in 1915, 8 years later after his thought experiment.

Here is another question, people ask every single time, are black holes true or are they fake, just science fiction? Actually, in the modern times, scientists, and data analysts prove that black holes are true, sending radio signals from outer space to Earth, similar to the Jocelyn Bell quasar discovery story. Black holes are also involved in Einstein's general theory of relativity, and Einstein proves to everyone that black holes will form one day in reality, because they are created of gravitational pulls. In 1916, Karl Schwarzchild also proved that black holes have a central region that the gravitational pull it exerts is so powerful, that light can't even escape, or any kind of matter can't escape.

So, what did Einstein theorize about black holes in his general theory of relativity, and its connection to gravity? In his general theory of relativity, it proves that gravity is not a normal force that pulls two objects towards each other, but as the fabric of time and space warping massive objects, and small objects around the massive object, and would later form an event horizon, also known as the "point of no return" that is now a black hole's inner section, that exerts a powerful gravitational pull so strong, that light and any kind of matter cannot escape.

But the actual proof of a black hole can exert such a strong gravitational pull, that light and any matter can't even escape from the event horizon was actually made by John Michell, an English county parson in 1783. Here is another question, that other people ask, what is the fourth dimension, and why is the spacetime curvature four-dimensional? Other people think that 4D does not exist, and does not exist in 4D video games, and they think there is no such thing as 4D video games and 4D objects in reality, but they are only 3D ones, well, it is only their own beliefs. But did scientists prove that 4D is real? Well, it actually depends in dimensional geometry, *which I explained earlier on Chapter 2.1 page 27*. Ok, 4D does exist,

Everything four-dimensional exists, 4D video games like Miegakure, and 4D Golf do exist, even 4D films exist as well. Scientists proved that 4D is real and does exist as well, obviously. Here is an image at the bottom right of the page, the four-dimensional cube itself, the terrific tesseract.

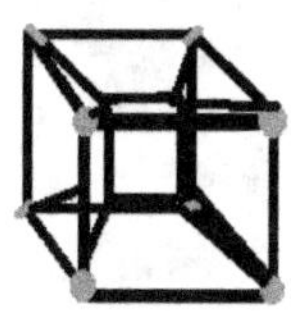

What is the fourth dimension, exactly, and why did Einstein develop a concept of the spacetime four-dimensional? But how did he know that the spacetime aka gravity four-dimensional?

So, the fourth dimension is actually time, which is connected to three dimensions of space, length, width and height, to form the spacetime curvature like Einstein explained in his general theory of relativity. This four-dimensional theory is one of the best explanations Einstein has proven so far in the past, but wasn't proven in reality, but only in scientifically hypothetical facts.

What is gravitational lensing in Einstein's general theory of relativity? Gravitational lensing is a type of cosmic phenomena that happens when light crosses along the massive object in the spacetime curvature, it bends, causing a distorted and reflective images of faraway or distant objects, this type of lensing is also found on black holes, light years away from our solar system, gravitational lensing also proves that the object has a powerful gravitational force that pulls objects, but will not escape from the black hole's strong gravity.

Here is the image below showing the black hole and its gravitational lens in space:

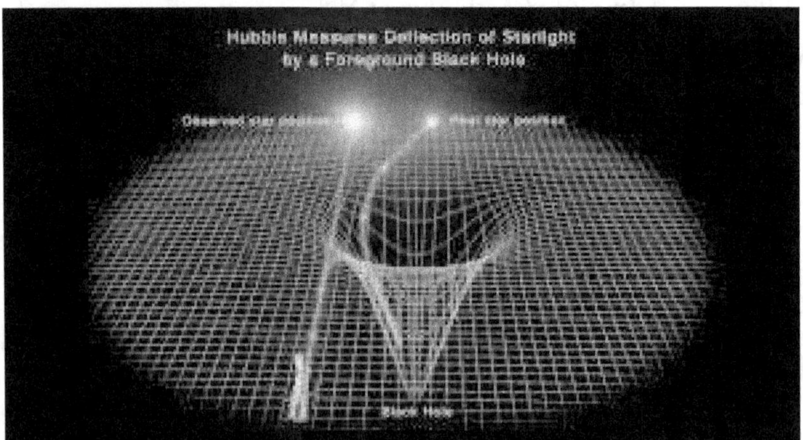

What is more to gravity in Einstein's general theory of relativity? Gravitational lensing, gravitational time dilation and also gravitational redshift of black holes. You may wonder what is gravitational time dilation, and the gravitational redshift of black holes? Well, in his general theory of relativity, Einstein proved that gravitational time dilation is actually a phenomenon where the time slows down when an object exerts a powerful and stronger gravitational pull.

For example: When a clock is at a higher altitude, it ticks slower, because of strong gravity, but when a clock is at a lower altitude, it ticks faster, because of weak gravity. This was predicted by Einstein in his general theory. But here is another question, what is gravitational redshift of a black hole, anyway?

Well, gravitational redshift is a phenomenon in Einstein's general theory of relativity that happens when light loses energy, causing the light to decrease its frequency and increase its wavelength, causing a redshift. You may also wonder why do scientists call it a redshift? The reason it is called redshift, it is because in the electromagnetic spectrum, when the frequency of light radiation decreases, the wavelength increases, which leads to the reddish part of the electromagnetic spectrum (infrared rays).

Well, frequency and wavelength are inverse in the electromagnetic spectrum. When frequency increases, the wavelength decreases, and when the frequency decreases, the wavelength increases during the gravitational redshift.

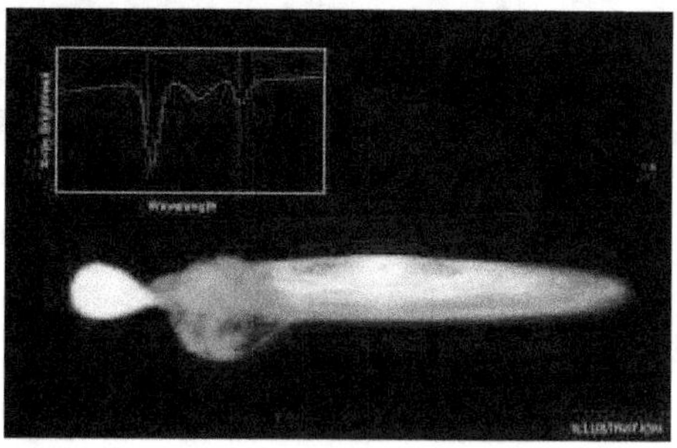

CHAPTER 8.2

PHYSICS & THE SPECIAL THEORY OF RELATIVITY

WHAT DOES E = MC² MEAN, AND ARE ALL LAWS OF PHYSICS ACTUALLY THE SAME?

Have you ever wondered what does $E = mc^2$ mean, exactly? In this sub-chapter, we are discovering more about Einstein's special theory of relativity, and how Einstein explained it in the early 20th century, and why the mass-energy equivalence formula is so renowned to the world.

What is Einstein's special theory of relativity? Einstein's special theory of relativity actually explains the similarity and the relationship between mass, space, time, energy and the the whole laws of physics are entirely the same for non-accelerating observers, and the speed of light is constant for all observers, and proves that mass and energy are equal as well, as it shows on the mass-energy equivalence formula: $E = mc^2$ (E for energy), (m for mass), and (c^2 for speed of light squared). The theory was published in 1905, 120 years ago, in his book English-titled "*On the Electrodynamics of Moving Bodies*", and German-titled "*Zur Elektrodynamik bewegter Körper*" by Albert Einstein.

Okay, so what does it mean by "non-accelerating observers"? Well, it actually means, that the laws of physics works the same for someone standing still as they do for someone moving at a continuous speed. And what does it actually mean by "the speed of light is constant to all observers"? It also means that all observers measure the speed of light in a vacuum (no space) at approximately 299,792.458 kilometers per second. So what does *$E = mc^2$* actually mean, exactly? The most popular mass-energy equivalence formula explained by Einstein in 1905, it means that small amount of mass contains a huge amount of energy. As the speed of light is the ultimate limit, that proves that matter can never reach the speed of light.

You may also wonder why doesn't matter reach faster than the speed of light? Well, the reason matter doesn't reach faster than the speed of light, is because the object's mass increases, and its speed increases as well, but in order to reach the speed of light, the object's mass and energy require to accelerate, and both of the mass and energy become infinite, creating infinite energy and infinite mass, in order to reach faster than the speed of light.

$E = mc^2$, also known scientifically as the mass-energy equivalence equation, why is this equation so interesting, and why is it the most known and popular equation in the world, why? In Einstein's special theory of relativity, the object's mass increases, and its energy increases as well, but in order to reach the speed of light, both its mass and energy must accelerate and require both infinite energy and mass and that is all the work of $E = mc^2$, matter cannot reach the speed of light, nothing can beat the speed of light, actually nothing is faster than light.

But light has some terrible consequence and an opposition, firstly its consequence, it cannot escape the black hole's event horizon, and its opposition, is obviously darkness.

But in reality, physics has proven once again that light is faster than anything on earth and the entire universe or multiverse, but in fiction, DC Comics' fictional character the Flash is faster than the speed of light. *Discover more about light and its connection with speed on Chapter 2.5, starting from page 49.*

So, why is $E = mc^2$ so popular anyway? Well, $E = mc^2$ is a very popular equation, it appears on animated series, social media and anywhere, why does it appear on those type of media, because it is a very simple, and a memorable equation that any person can remember.

But not only professional scientists like Brian Cox, Hakeem Oluseyi, Albert Einstein and other professionals out there, but the whole wide world knows the equation, because of its simplicity. $E = mc^2$ is one of the most renowned equations in the history of physics, ever explained by Professor Albert Einstein in his 1905 book "*On the Electrodynamics of Moving Bodies*" about the special theory of relativity.

There is also time dilation and length contraction in Einstein's special theory of relativity, what do those mean, anyway? Time dilation was explained in a different way on Einstein's general theory of relativity, that when the gravity is strong, the time slows down, or when the gravity is weak, the time speeds up.

Example: When the clock ticks slower, when a person falls from the first floor to the safety equipment like a jumping castle on the ground at a faster rate, and when the clock ticks faster, when a stone orbits around a huge rock slowly, showing that the gravitational pull is weak in the spacetime curvature.

So, what is time dilation on Einstein's special theory of relativity differ from Einstein's general theory of relativity? The time dilation in Einstein's special theory explains when the time passes more slowly when the object is at a high speed rate. As length contraction however, explains that when objects move at a faster rate, they appear shorter through the direction of motion, and when objects move at a slower rate, they appear longer and vice versa.

So, what is an observer in the special theory of relativity? An observer is defined as a reference frame or a grid that coordinates points to measure in the spacetime, measuring amounts of speed, mass and energy.

Here is another question that people ask daily: What is the difference between the general theory of relativity and the special theory of relativity, and sometimes called the theory of relativity? Modern people find it confusing of Einstein's theories in physics.

Let's start with the general theory of relativity:

- Einstein's general theory of relativity is the theory explained by Einstein about black holes and the spacetime curvature, and all about gravitational pulls, what happens with time when the gravity is strong or weak?
- Einstein explained on his general relativity theory that gravity isn't a normal force, but the spacetime curvature itself.
- Why? Massive objects warp the fabric of space and time, and smaller objects follow the path and are being pulled gravitationally around the massive objects, forcing the smaller objects to orbit around the massive object.
- Similar to planets orbiting around the Sun, how big or small they are, with different distances from the Sun, and different revolutions like Earth which is 15×10^6 km away from the Sun or an astronomical unit, revolves around the Sun for a year.
- While Neptune the eighth planet revolves for 164 Earth years, that will take years to do one revolution, because it is afar from the Sun.

- In another statement in Einstein's general theory of relativity, he also explained that a black hole has a gravitational pull so strong and powerful, that many celestial objects can enter the event horizon and never escape, not even light, he took that statement from English county parson John Michell, who wrote in his book in 1784.
- Time dilation of the general theory of relativity explains time slows down when the gravitational pull is stronger, and time speeds up when the gravitational pull is weaker.
- The gravitational redshift explains when light loses energy and its frequency decreases and wavelength increases, leading the electromagnetic spectrum to the red color.
- In the general theory of relativity, it also proves that the spacetime curvature, is not unidimensional nor 2D, 3D but four-dimensional, the fourth dimension is created when three normal spatial dimensions, length, width and height are connected with the non-spatial dimension, time to form a four-dimensional spacetime curvature.

Here is an image below of the spacetime curvature, the fourth dimensional geometrical property in the universe.

- Finally, what we have learnt about Einstein's general relativity theory, is gravitational lensing, found on black holes and the spacetime curvature as well, it is a cosmic phenomena, that when light crosses its massive object in the spacetime curvature full of strong gravitational fields, it bends and creates a distorted image of the faraway distant objects in space.

Let's continue with Einstein's special theory of relativity:

- Einstein's special relativity theory explains the relationship between mass, energy and other laws of physics and their similarity, and the speed of light remaining constant to all observers.
- It also explains that the laws and physics are identical and the same for non-accelerating observers, for objects who do not move at high speeds. While the speed of light is constant to all observers, for objects that move at high speeds.
- The mass-energy equivalence equation "$E = mc^2$" was shown on Einstein's special relativity theory, and it is one of most popular, unforgettable and simplest equation in the world.
- The $E = mc^2$ equation means that energy and mass are equivalent, but matter or mass will never reach the speed of light, as it will require both infinite energy and infinite mass to reach the speed of light. As for limited energy and mass, the speed of light is unreachable. Here is the image at the bottom right of the previous page showing the popular equation in the world.

Finally, what is the theory of relativity itself? The theory of relativity is a combined physics theory shown by Albert Einstein containing two theories, the special theory of relativity and the general theory of relativity.

Well, we learnt how the theories of relativity (the special and the general) work, and what they mean, and one of the most important and interesting questions so far; what does $E = mc^2$ mean, exactly?

The most mind-boggling and fascinating thing we have discovered in the general theory of relativity is that gravity is notjust a force that pulls objects together, but a geometrical property that is four-dimensional and has the fabric of space and time.

Even in the special theory of relativity of what the equation $E = mc^2$ actually means, that energy and mass are equivalent, and the speed of light is constant to all observers, but in order to reach the speed of light, it requires infinite energy and mass to reach the speed of light.

Physics is so incredibly fascinating about Einstein's mind-boggling theories of relativity. Albert Einstein once said "*Weak people revenge, Strong people forgive, Intelligent people ignore*". Here is the image at the bottom right of the page, the incredible professor and the big brain guy himself, Professor Albert Einstein (1879-1955), the "theory of relativity" mastermind.

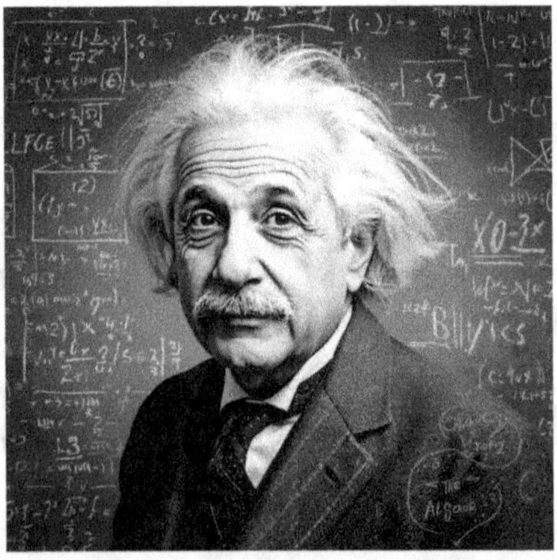

CHAPTER 8.3

PHYSICS & THE THEORY OF EVERYTHING

IS IT REALLY THE THEORY OF LITERALLY EVERYTHING AND CONNECTION TO OUR ENTIRE UNIVERSE?

Have you wondered there is a theory called the theory of everything, where it has many hidden secrets about our universe, maybe? In this sub-chapter, we are discovering more about the ToE, or the theory of everything. Some people say that the theory of everything, is the unsolved theory that no scientist can solve, but they are others who are attempting to solve this confusing theory.

But here is a question: why is it called the theory of everything? If it is the theory of everything, is it the theory that has many hidden secrets in our universe, from quantum mechanics to general relativity and all of laws of physics. What is the theory of everything, exactly?

The theory of everything is the theoretical framework that contains all physics theories into one huge, unsolved and confusing theory. The ToE is also called the final theory, meaning it is the unsolvable theory, from Newton's laws of gravitation to Einstein's theory of general relativity. The theory of everything contains unifications of four fundamental forces of nature that are linked to this confusing theory, like electricity, magnetism, weak nuclear forces and gravity.

But, why is it called the theory of everything? Well, it is called "the theory of everything" because the objective the scientists are trying to do, is to unify all the fundamental forces of nature in our universe into one single theory, explaining the everything of these forces.

The theory of everything also explains how the big, giant and macroscopic world (e. g. galaxies, planets, black holes, stars and celestial objects) from Einstein's general theory of relativity is connected to the small, tiny and microscopic world (e. g. quarks, electrons, protons, leptons, neutrons and atoms) from the quantum theory. It is also called "the theory of everything" to conceptualize how all the forces of nature in the universe are going to be explained into one single unified theory.

Here is the Venn diagram image below, created by Wikipedia user, Cmglee, meaningfully describing how the theory of "everything" looks like and what does it actually mean:

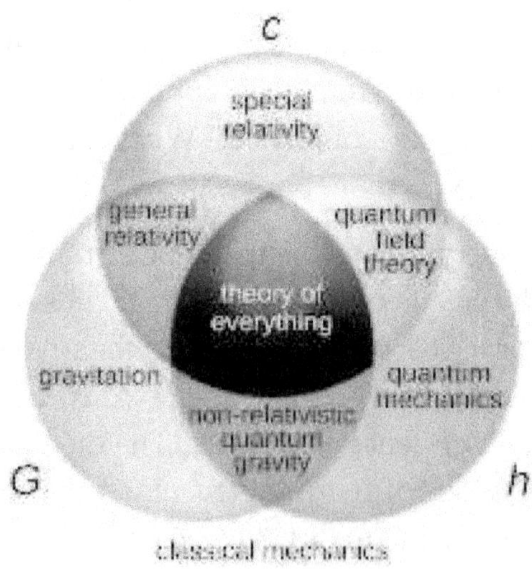

You may wonder what does the *c*, *G* and *h* mean on the Venn diagram image you saw on the previous page?

- The *c* represents of the speed of light in a vacuum, the constant of Einstein's special theory of relativity.
- The *G* represents Newton's gravitational constant, the constant of both Einstein's general theory of relativity and gravitation or gravity.
- The *h* represents the Planck constant, the constant of quantum mechanics. The Planck constant is named after German physicist Max Ernst Ludwig Planck.
- What is the overlap or interaction between the three constants in physics?
- The overlap of *c* and *G* (the speed of light and Newton's gravitational constant) represents Einstein's general theory of relativity.
- The overlap of *c* and *h* (the speed of light and the Planck constant) represents relativistic quantum field theory, it is a theoretical framework that combines quantum mechanics, general relativity, and fields to explains the subatomic particles (e. g. electrons) and their interactions.
- The overlap of *G* and *h* (Newton's gravitational constant and the Planck constant) represents non-relativistic quantum gravity.
- Non-relativistic quantum gravity is a type of theory that combines quantum mechanics and a non-relativistic version of gravity, describing that velocities are less than the speed of light and the gravitational field is weak, and it is also the opposite of the theory of relativity.

The theory of everything, the most unsolvable problem in the world of physics. Physicists are attempting to find out and explain the "everything" the laws of physics has, into one single unified theory. But when will they find out this complicating, confusing, difficult and unsolvable theory, and when will they explain? Well, it looks like it is not quite sure, when physicists will solve this difficult theory of "everything". But the theory of everything is not the only unsolved problem in physics.

There are other unsolved problems out there in physics, that are more difficult to solve:

- Starting with, Dimensionless physical constant, it is a physical constant that has no dimensional quantities or dimensionless to be in fact, for example a pure number like one has no units attached, which is a complicating and confusing unsolved problem in physics as well.
- The black hole information paradox, is the unsolved problem in physics, that focuses on the predictions of quantum mechanics and general relativiry are combined. Einstein's theory of general relativity predicts the black holes' existence that is on the region of the spacetime from which nothing not even light can escape.
- Quantum gravity is a type of gravitational field in theoretical physics, that focuses on the unification of quantum mechanics and gravitational theory.

Well, in physics and the world of problems, physicists and other scientists who study different fields, are used to solving problems. But those that are hard and difficult like the theory of everything, takes years to attempt to solve this complicating theory. But the theory of everything will remain the most unsolved physics problem in the history of the universe.

CHAPTER 8.4

PHYSICS & THE BIG BANG THEORY

HOW DID THIS CATASTROPHIC EXPLOSION IN HISTORY, CREATE THE UNIVERSE AND LED TO EXPANSION UNTIL TODAY?

The Big Bang! One of the most incredible, mind-boggling, and fascinating explosions in the entire universe, that led to creation. But people ask: why are we here? What made us? Who made us? Why is there so many life? Well, this is a debate between theoretical physics and Christianity belief. We, humans believe that the Big Bang created the universe, how and why? This is the sub-chapter to all your curious answers.

On our own Christian beliefs, I mean In the Holy Bible, the book of Genesis, it focuses on the Creation of our World, the Book says, In the beginning, God, our Father, created the heavens and the earth. On the first day, He made the heavens, and the earth, and said “Let there be light”.

On the second day, of the Creation, He spoke and created the sky and the sea, and separated the waters above the earth, forming a sky creating a firmament and the water on the earth as well.

The third day, He created dry land and plants, then on the fourth day of Creation, He created the Sun, moon and the stars, then on the fifth day of the Creation, the water and the sky animals were made by Him, on the sixth day of the Creation, the land animals and us humans were created. The last and seventh day of the Creation, He rested from all the work He had done, and blessed the seventh day or Sunday, the holy day until today.

But in theoretical physics, what is the Big Bang and its theory? The Big Bang is a belief of what scientists prove that it is the biggest and catastrophic explosions ever created the universe in history. But who theorized the Big Bang first? This Big Bang theory was first credited by Belgian cosmologist and Catholic priest, Georges Lemaître in 1931, who was doing a hypothesis of the primeval atom, which led to the universe's creation, he theorized about in 1931, 94 years ago.

You may also wonder what exactly the primeval atom is? Lemaître stated that the universe was initially made by a single particle which he called the primeval atom, it is a particle that disintegrated in the Big Bang explosion, and created spacetime and led to the expansion of the universe until today. The primeval atom exploded because of nuclear fusion reactions during the Big Bang explosion. Many physicists believe that the Big Bang was the cause of the universe's origin 13.8 billion years ago.

But there is one physicist who rejected and didn't believe the Big Bang theory and it is English physicist Sir Fred Hoyle. The reason why rejected because he thought there was our creator, Father God, from the Holy Bible, but he was an atheist, meaning he didn't believe in God and the Big Bang as well.

You may wonder how was the universe exactly created? So, in the Big Bang theory, when the explosion occurred, the universe started as an extremely hot, extremely small, and an extremely dense spherical ball of heat, from then until today, it has expanded and expanded, and then it has become less dense and cooler. So, what was there before the Big Bang explosion? That is a mind-boggling question, that scientists and the curious are interested in finding out. Well, 13.8 billion years ago, the universe was just a singularity at first, the point smaller than a subatomic particle.

So, what does it mean by a point smaller than a subatomic particle? It means that the singularity is the point in the fabric of space and time or in the spacetime curvature that has infinite density, but has no volume.

The singularity of the Big Bang theory was first credited by British theoretical physicists, Sir Stephen Hawking and Roger Penrose who developed the Penrose-Hawking singularity theorems using black holes and cosmological models, demonstrating that the spacetime of an expanding universe must end with the singularity from the past, which is the Big Bang singularity. What does that actually mean, by the universe's expansion must end with the past's Big Bang singularity?

So, it actually means, the universe will be reversing its expansion or contracting, reversing back to the singularity point, you may also wonder will time reverse too? No, actually, time won't reverse, but it will keep going forward, while the universe contracts after its expansion ends in the future. The reversal version of the Big Bang is called the Big Crunch, termed by Hawking and Penrose on their demonstration.

Here is another interesting question, will there be another Big Bang? Well, it is not quite sure whether there will be a second Big Bang, but physicists and cosmologists believe that there will be a second Big Bang after the future contraction of the universe, if it will happen. It is not just that, it is believed that the universe will expand, then contracts again and again, there will be a repeated cycle of many Big Bangs, it will explode over and over again.

In 1964, American physicist Arno Allan Penzias and radio astronomer Robert Woodlow Wilson discovered the CMB which stands for cosmic microwave background, estimating its temperature at 3.5 K, when they conducted an experiment with the Holmdel Horn Antenna, which is a large microwave horn antenna used as a satellite communication device and a radio telescope that was used in the 1960s. How did they discover the CMB? Arno and Robert discovered the CMB using their antenna, by observing the cosmic microwave background radiation from outer space, and marks the biggest cosmological discovery in the world of physics and cosmology.

So, you may wonder what connection does the cosmic microwave background radiation actually have with the Big Bang theory? Well, the CMBR or cosmic microwave background radiation was created 380,000 years after the catastrophic Big Bang. So, what is the CMB, exactly? The cosmic microwave background is actually a cooled portion of light that traveled freely, leading to the infinite expansion of the universe.

Physicists and cosmologists believe that the CMB has evidence that the Big Bang existed, because its light is being stretched and the cause of the Universe's expansion forever. The CMBR is not just light that has small amounts of its radiation in one single cosmic area, but the entire universe, since it originated 380,000 years after the Big Bang. The CMB is also known as the afterglow light pattern.

Here is the image below, containing a source of information that shows the Big Bang and the cosmic evolution timeline.

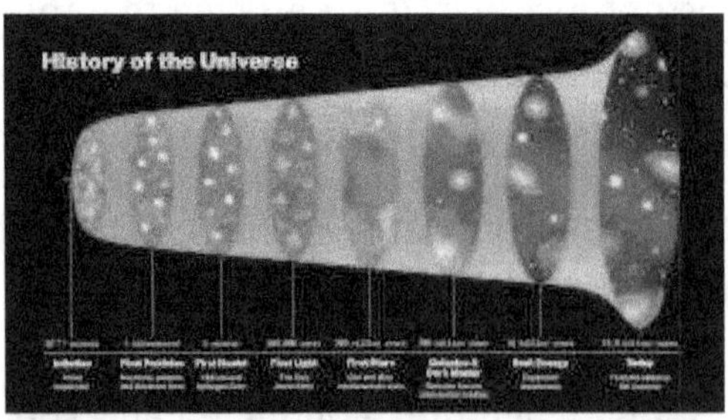

So, what is more proof or evidence that the Big Bang actually exists?

- In 1929, American astronomer Edwin Hubble clearly stated that the galaxies are moving away from us, when the universe continuously expands.
- The cosmic microwave background radiation (CMBR) the radiation that is filled around the entire observable universe.
- The galaxies are redshifted, the galaxies' light radiation, decreases its frequency and increases its wavelength, shifting to longer and redder wavelengths.
- The steady-state theory, which states that the universe doesn't change when expanding, but remains to its original, normal state steadily.
- The oscillating universe theory, which is the reversal version of the Big Bang theory, which states that if the universe's expansion ends, it will reverse its expansion, or it will contract in other words, and reverse back to the singularity point of where the Big Bang occurred, then explode once again, and create another universe, and re-expand, then re-contract, and then re-explode again and again, and never stops.
- The abundance of different elements like hydrogen and helium took part in the explosion of the Big Bang from an extremely hot and extremely denser phase to the cooler phase.

The Big Bang theory is one of the most agreeable theories in the universe, by mostly physicists, non-physicists and cosmologists, except English astrophysicist Sir Fred Hoyle, himself. But according to the holy book of Genesis, it says the universe was a ball of energy and light, it is also described the same topic on the Big Bang theory.

But wait a second, how did the Big Bang create matter? During the Big Bang, a few millionths (1/1000000) of a seconds or microseconds later, quarks produced protons and neutrons and other subatomic particles. Then minutes later, those protons and neutrons combined into nuclei. As the universe began to expand and cool its temperature, and it took 380,000 years for electrons to be trapped in orbit around a nuclei like a single atom works, and it was also the same event where the CMBR appeared.

Then 200 million years later after the Big Bang, the first ever stars formed in clouds of immense gas in space called nebulae. Stronger and heavier atoms like carbon, oxygen and iron are being produced inside the structure of stars, until they die and explode in powerful stellar explosions called supernovae.

Research shows that 27% of the universe is filled with dark matter, a type of matter that does not emit any light or any electromagnetic radiation (e. g., ultraviolet radiation), and it is invisible to the human eye, and 68% of the universe it is filled with dark energy, it is a mysterious type of energy that is filled almost the whole universe, and causes the universe to accelerate its expansion rapidly, and also invisible.

While five percent of the universe contains visible matter, a type of matter that is visible to the human eye like light, the Sun, the Moon, galaxies, stars, black holes, and celestial objects that are out there in outer space.

So you may wonder how did dark matter, dark energy and visible matter form during the Big Bang?

Who discovered dark matter? Dark matter was discovered by Swiss-born astronomer Fritz Zwicky in 1930. Zwicky published a paper in which he observed studying the Coma Cluster, which is a cluster of galaxies. He observed and saw the galaxies moved rapidly in the cluster by the gravity created by some peculiar matter. He wondered what kind of thing causes the galaxies to stay together.

He knew that there were some kind of invisible matter that causes the gravity to hold the galaxies together, and he called that type of matter "dunkle materie" in German, but was later translated in English as dark matter. Dark matter, doesn't absorb, interact and emit light, but its gravity bends light, that phenomenon is called gravitational lensing, the same phenomenon found on black holes.

What is the difference between dark matter and dark energy? You may also wonder why do they have the word "dark"? These cosmic phenomenons use the word "dark" because it means they don't absorb, interact and emit light, but they are also invisible.

So what is the difference between dark matter and dark energy?

- Dark matter is a type of matter in physics that does not absorb, interact and emit light and contains 27% in the universe.
- While dark energy on the other hand, it is a mysterious form of energy that contains 76% in the universe, and it is spread throughout the universe at larger scales.
- Dark matter and dark energy with galaxies are like magnets but with opposite abilities, for example: dark

matter attracts with galaxies meaning it pulls galaxies together, while dark energy on the other hand, repels with galaxies, meaning it pushes galaxies apart from each other.

- Then, moving to visible matter, what exactly is visible matter? Visible matter is the type of matter that is perceivable to the human eye, like the Sun and other types of visible matter we know.

But in astrophysics, there is the another type of matter in the universe, that is opposable to matter. Antimatter, itself. What is antimatter? Antimatter is a type of matter that is opposable to normal matter. Antimatter can be referred as the most dangerous type of matter ever existed on the universe. Why is it referred as the “dangerous” type of matter?

Antimatter is referred because it can destroy matter in an annihilative way, like two empires having a war to see which the empire is the strongest. But on earth, antimatter exists, do you know that antimatter is being produced on bananas? Yes, it is true, bananas are just ordinary fruits that produce potassium. So does potassium contain antimatter? Yes, potassium as well contains antimatter. But, this is not a warning of stopping eating bananas, you can eat bananas since they are healthy for the human body.

So what have we learnt about the early development of our universe? The Big Bang theory has some evidence that many physicists have proven to us for many decades, like Hubble’s 1929 statement of galaxies moving away from us, and the universe’s redshift, dark matter, dark energy and visible matter as well and why they are involved in our universe. The Big Bang theory is one of the most fascinating and incredible theories ever stated in physics, thanks to the genius himself, Georges Lemaître (1894-1966).

CHAPTER 8.5

PHYSICS & THE STRING THEORY

IS THERE REALLY A TENTH DIMENSION?

Have you ever wondered how many dimensions we have in our universe in the fabric of space and time? How many dimensions you know we have in our universe? 2? 3? 4?

Well, we people think that dimensions end at 3D or 4D, but in physics, dimensions don't just end at 3D or 4D, but infinitely, maybe at 100D or more, but mathematicians and physicists haven't found out about 100D or more dimensional properties in geometry and physics. There is this theory called the string theory in physics. What exactly is the string theory?

The string theory is a theoretical framework in physics that uses point-like particles being replaced by one-dimensional particles called strings. So, does it mean unidimensional particles aren't point-like but wavy-structured like normal strings?

Well, on the string theory, physicists theorize that point-like particles, are not just ordinary points with no spatial extent, but they are wave-like particles that vibrate on the theory. But there was no experiment to prove that the string theory exists as the most fundamental theory in physics and nature as well.

The string theory is similar to the theory of everything, because scientists are trying to unify the theory of quantum mechanics and Einstein's general theory of relativity, while the theory of everything is being attempted to unify all fundamental forces of nature and physics, electricity, magnetism, strong nuclear forces and weak nuclear forces.

In theoretical physics, the string theory is also called the superstring theory, because it is the theory that states that the universe is 10-dimensional, but we humans only observe the universe as a (3+1=4)-dimensional spacetime area.

Three spatial dimensions are length, width and height while the fourth dimension is theoretically, time stated on Einstein's general theory of relativity focusing on the spacetime curvature and gravity. When the three spatial dimensions merge with the singular dimension of time, they form a spacetime, explained on Einstein's general theory of relativity as well.

So you may wonder we have ten dimensions in our universe, but we know that four dimensions exist, but what about the remaining six dimensions, what do they actually mean? The remaining six dimensions in our universe are folded and compactified very tightly, and us humans cannot perceive those remaining six dimensions clearly.

There is a scientific model used in physics to observe six-dimensional compactification with the four-dimensional spacetime in our universe, the Calabi-Yau manifold. What exactly is the Calabi-Yau manifold and why is it important in the superstring theory in theoretical physics?

The Calabi-Yau manifold is a compactified six-dimensional and a Ricci-flat manifold that involved in the superstring theory. Here is the image bottom right of the page of the K3-type Calabi-Yau manifold, created by Wikipedia user, Andrew J. Hanson.

What does it actually mean when the remaining six dimensions of the universe, are compactified and what is a Ricci-flat manifold? Firstly, a Ricci-flat manifold is a type of manifold that zero Ricci curvature, meaning, the matter and energy would converge or diverge in the spacetime is balanced out. A Ricci curvature is a geometric quantity that measures the volume of small ball of points that changes and being transported through geodesic models.

Secondly, what does it mean by compactification of the remaining six dimensions in our universe? Compactification of the remaining six dimensions in our universe are actually extra dimensions predicted by the superstring theory that states that these extra dimensions are not infinite and observable, but they are curled up into tiny, microscopic space that is too small for us to see exactly.

But there are other types of theories similar to the string theory focusing on dimensional geometry:

- The bosonic string theory, which is the first version of the string theory or superstring theory, explains that the fundamental particles are unidimensional rather than point-like particles which are dimensionless, it requires 26 spacetime dimensions at first.
- The superstring theory, which is the modern version of the string theory, explains that the point-like particles are being replaced by unidimensional particles called strings, which have the ability to vibrate, it requires six dimensions thar cannot be perceived, and four spacetime dimensions, which merge to 10 dimensions, but it is unproven same to the bosonic string theory.

- The M-theory, is the theoretical framework in physics that will unify all fundamental forces and particles like strings and other objects called “branes”.

So, let’s discover the history of the string theory and the Calabi-Yau manifold? What was the string theory before becoming the superstring theory? I mean, what was the first version of it, and who theorized this incredible and fascinating theory in the world of physics?

The history of the string theory originated in the 1960s, in the year 1968, Physicist Gabriele Veneziano discovered the Euler beta function describing the strong nuclear force, it is known as the Veneziano model, then later, a group of scientists named Leonard Susskind, Holger Nielsen and Yoichiro Nambu explained that the one-dimensional particles are also vibrating strings, and from that moment, the theory became known as the string theory until today. Then, supersymmetry was discovered by physicist Pierre Ramond in 1974, which led to the superstring theory, which has a connection with fermions (half-integer spin) and bosons (integer spin).

Supersymmetry is a theoretical framework in physics that predicts the existence of symmetry between bosons (particles with integer spin) and fermions (particles with half-integer spin).

The Calabi-Yau manifold’s early development and evidence dates back to 1958, where Italian-American mathematician Eugenio Calabi guessed that there is a complex Kähler manifold with a first Chern class can have a Ricci-flat curvature. Then later in 1974, Chinese-American mathematician, Shing-Tung Yau, proved Calabi’s estimation and demonstrated that the Kähler manifold can have a Ricci-flat curvature. Then later in 1985, a group of scientists, Philip Candelas, Gary Horowitz, Andrew Strominger and Edward Witten coined that kind of manifold, the Calabi-Yau manifold, representing the honour of Calabi and Yau's works, from Calabi's estimation to Yau's evidence of proving that the complex Kähler manifold does have a Ricci-flat curvature.

What was the string theory before becoming the superstring theory? The original version of the string theory, was the bosonic string theory. The bosonic string theory was first developed by physicist Gabriele Veneziano, the bosonic string theory states that the fundamental particles are described as unidimensional vibrating strings, and it requires 26 spacetime dimensions.

What does the "bosonic" word mean in the bosonic string theory? The bosonic word actually means that the bosonic string theory includes, elementary particles that have an integer spin called bosons, meaning they require 26 spacetime dimensions in our universe. But why so many dimensions? 26?

The bosonic string theory requires 26 spacetime dimensions because it was needed to cancel the Weyl anomaly, is a quantum phenomenon that would lead to negative-norm states and faster-than-light particles like tachyons. But what are tachyons?

Tachyons are hypothetical particles that are believed to be faster than light, but they do not exist, and physicists haven't shown any evidence of the true existence of these tachyons. Here is the image of these hypothetical tachyons at the bottom right side of the page, in theory, they are being stated as particles that are faster than the speed of light itself.

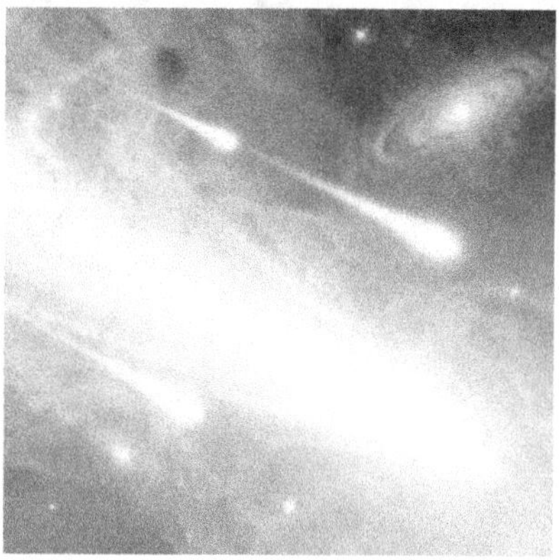

So, how did the bosonic string theory evolve to the presently known superstring theory until today? The reason was supersymmetry, which has developed a link between both elementary particles, bosons (with integer spin) and fermions (with half-integer spin).

While the previous bosonic theory, included only bosons, and the reason behind 26 spacetime dimensions. But the superstring theory, the modern version of the string theory, reduced the required spacetime dimensions from 26 to 10, (nine spatial dimensions plus one time dimension), which is a sensible part ofthe theory focusing on compactification in the four-dimensional universe.

Well, in the world of confusing and complicating physics, the string theory is similar to the theory of everything, many physicists are attempting to unify Einstein's general theory of relativity and quantum mechanics into one single theory. The string theory remains unproven until today, but physicists are still attempting, we will never know when the string theory, or the theory of everything will be proven true in the name of physics, maybe in the future, where more mathematical and geometrical dimensions will develop from 4D to 5D or more. Discover another theory similar to the string theory known as the M-theory on the next chapter, after this one.

Here is the image below showing the one-dimensional vibrating strings from the unproven string theory.

CHAPTER 8.6

PHYSICS & THE M-THEORY

WILL THE UNIFICATION OF THE STRING THEORY UNRAVEL THE HIDDEN SECRETS OF THE UNIVERSE AND ITS DIMENSIONS?

Have you ever wondered in geometrical physics, what unification or compactification actually are from the string theory? And have you also wondered if the string theory is finally proven, will it unravel the hidden secrets of the universe and its dimensions of space and time? Let's investigate and find all the answers of these fascinating and interesting questions.

So, what is the M-theory in physics, exactly? Is it unproven like the string theory and the theory of everything? The M-theory is a theoretical framework in physics that aims to unify Einstein's general theory of relativity and quantum mechanics, and it is believed that it has 11 dimensions, 10 spatial dimensions and 1 time dimension, similar to the string theory. It includes our four dimensions we have in our universe (three spatial dimensions plus one time dimension) plus seven dimensions that cannot be perceived and are curled up and shrunk in tiny microscopic space.

The M-theory is also a theoretical framework in physics that describes that the point-like particles are being replaced with unidimensional vibrating strings and higher-dimensional particles called branes.

So, what is a brane, exactly? A brane is a multidimensional object that can a be zero-dimensional point or a 0-brane, or a unidimensional string (1-brane), or a two-dimensional membrane

(2-brane) and it is part of the M-theory. There are different kinds of branes involved in the M-theory.

- P-branes: A different type of brane that has a number of spatial dimensions it adds, like 0-brane, a point, 1-brane, a string, (one of the most fundamental objects in the string theory), and 2-brane, a membrane or a 2D figure like a square.
- D-branes: A different type of brane where fundamental strings will end.
- Black branes: A different type of brane on Einstein's field equations that resemble black holes but are extended to some dimensions.
- M-branes: A different type of brane that is generalized further to include higher dimensional objects like the M2-brane (a two-dimensional brane), and the M4-brane (a four-dimensional brane).

What crucial role do branes play in the M-theory? Branes play an important role in the M-theory, because they are higher-dimensional objects that generalize strings and unifying all five superstring theories into a single 11-dimensional model in the universe. Here is the image below, showing the impossible 11-dimensional model from the M-theory.

You may also wonder what are all those five superstring theories? Buckle up, readers, it is about to get tough, and complicated as well. Those five superstring theories, physicists are attempting to unify into a single 11-dimensional model are listed below:

- The Type I superstring theory: a 10-dimensional theory that has unoriented open strings (which has endpoints) and closed strings (which forms loops), unoriented meaning there is no difference between the two possible directions along the string.
- The Type IIA superstring theory: a 10-dimensional theory that has oriented closed strings and has non-chirality (meaning left to right symmetry, superimposable mirrored image, two objects that are placed one over the other, like a piece of paper, it has two different sides).
- The Type IIB superstring theory: a 10-dimensional theory that has one-dimensional vibrating strings with both open and closed strings, and it is different from other superstring theories, because of the weird properties of its strings.
- The Type IIB also involves closed strings with asymmetrical vibrational patterns based on their travel direction and open strings that are connected to D-branes, which have the even number of dimensions.
- The Heterotic (E8 x E8) superstring theory: a 10-dimensional theory describing fundamental particles as one-dimensional vibrational strings, and it is known for its gauge group which (E8 x E8) and supersymmetry, and a relationship with two elementary particles, bosons and fermions.

- The Heterotic SO(32) superstring theory: a 10-dimensional theory describing fundamental particles as one-dimensional vibrational strings, this theory is identical to the Heterotic (E8 x E8), but the difference is the gauge groups.

We have now known the five superstring theories, the Type I, Type IIA, Type IIB, Heterotic (E8 x E8) and the Heterotic SO(32). You may wonder what exactly is a non-chiral object and what are gauge groups shown on the heterotic theories?

A non-chiral object in physics, is a different kind of object is also called a superimposable mirror imaged object, meaning that the sides of an object are left to right, but mirrored, and the objects are also placed one over the another, and are aligned perfectly.

Gauge groups like (E8 x E8) are set of mathematical symmetries that describe bow a theory remains unchanged even after transformations are made to the system's description. So, what do these gauge groups E8 x E8 and SO(32) mean in mathematics and in these heterotic superstring theories?

The gauge groups E8 x E8 and SO(32) mean that the related to a duality concept called T-duality, meaning that the heterotic E8 x E8 superstring theory is compactified on a circle of radius *R*, is actually equivalent to the other SO(32) superstring theory which is also compactified on a circle of radius *1/R*. Here is the image at the bottom right of the page showing the Type I string theory's exemplar objects.

So, who developed the M-theory, and why was it developed, exactly? Well, we date back to 1995, theoretical physicist Edward Witten developed the M-theory, because he wanted to unify all the fundamental forces like gravity and other theories that explains all the secrets of the universe.

The reason the M-theory was developed, because it acts as the unification of all fundamental forces of the universe, into one, single powerful and fundamental theory that could explain everything about our universe. Well, the main question of this sub-chapter is actually is: Will the unification of the string theory and the M-theory itself unravel the hidden secrets of the universe and its required dimensional properties or in simple form dimensions?

Well, maybe yes, but physicists around the world are still on theoretical research, if these theories are proven correct, they would unravel the secrets of the universe, like how the Big Bang led to the never-ending expansion of our universe, or proving the 10 (string theory) and 11 (M-theory) dimensions whether are compactified and curled up, and unobservable to us humans, and the other four (three spatial and one time) are observable to us.

You may wonder what do the words, compactification or unification mean? What do these words actually mean? Well, the word “compactification” was first introduced by Russian mathematician, Andrey Nikolayevich Tikhonov, and it also means, to make large quantity into a smaller quantity, and it is hard for us to see the small quantity, like atoms or other hidden dimensions out there in space, we humans cannot see even on Earth. The word "unification" was created by adding the suffix "-fication" to the verb "unify," which comes from the Late Latin "unificare" (to make one). The first recorded use of "unification" is from the 1840s, from the prefix “uni-” which means one in Latin. The word “unification” also means to combine all things into one single thing, like physicists are trying to prove the M-theory, string theory and the theory of everything, to unify all forces of physics into one powerful theory, that explains everything in the universe, if proven correct. But let's end the topic, and move to the question: What have we actually learnt from these theories, anyway? Let's start from Einstein's general theory of relativity, Einstein theorized in his book "On the Electrodynamics of Moving Bodies" published in 1905, that gravity is not just an ordinary force of physics, but a theoretical framework in his general theory called the spacetime curvature, that any large object can warp the fabric of space and time, and the smaller objects can move around the large object, forming a gravitational force, which is revealed in the theory as the spacetime curvature, e. g. the Solar System, proving the Sun is at the center, then all eight planets revolve around the Sun.

Then, Einstein's special theory focuses on the "$E = mc^2$" formula that actually proves that mass and energy are similar and are also equivalent, but the speed of light is different from mass and energy, meaning that the speed of light is unreachable, if the matter has finite and non-accelerating energy (unable to change speed, slows down).

But if the mass or matter has infinite energy, it can impossibly reach the speed of light. What we also learnt in the superstring theory, that the universe has 10 dimensions (nine spatial dimensions, one time dimension, while the six is curled up and compactified, and unobservable, three is spatial, and the one is time dimensional), similar to the M-theory which has 11 required dimensions, and the theory of everything is the unsolvable theory in physics attempting to unify all into one theory.

CHAPTER 9

EVERYTHING YOU NEED TO KNOW ABOUT PHYSICS & ITS LAWS

Have you ever wondered what laws of physics are? Laws of physics are actually fundamental principles that describe how our universe works, what does motion, energy, speed, force, and other forces of nature do in our everyday life, and why is physics involved in our everyday life as well, why is it crucial?

In this chapter, you will discover other laws of physics and their interactions with our modern world:

- What Newton's Laws of Motion are, and why is it the reason behind Earth's revolution around the Sun, not just the Earth, but the entire Solar System.
- What Kepler's Laws of Planetary Motion are, and why is it the reason behind the planets' motion from the cause of gravitational force.
- What the Laws of Thermodynamics are as well, and the fundamental workings of energy, and its connection with heat, light and even water, and renewable sources.

So, why are the laws of physics so important in our life? Well, the laws of physics are so crucial, because we humans understand these fundamental laws, and use them as the key tool to develop advanced technology like MRI scanners on hospitals, helping patients, and the laws of thermodynamics help with energy transfer, helping renew energy from nature like solar rays from the Sun and so on.

CHAPTER 9.1

PHYSICS & NEWTON'S LAWS OF MOTION

WHAT MAKES THE EARTH REVOLVE AROUND THE SUN WITH THE HELP OF NEWTON'S LAWS?

Have you ever wondered in your mind what does the phrase "What goes up, must come down" actually mean? You may wonder, the fictional character SpongeBob also said the same phrase, but what does it actually mean and why is this phrase everywhere I hear, and everywhere I see? Does this phrase have a connection with gravity and Newton's laws of motion?

Yes, the "What goes up, must come down" phrase has a connection with gravity and Newton's laws of motion. "What goes up, must come down" actually means, that it resembles gravity and how it works (for example: a person throws an object up, then the object begins to fall downwards, when it went upward). It does have a connection with Newton's laws of motion.

What are Newton's laws of motion, exactly? So, let's leave the linguistic side and move to the factual side, Newton's laws of motion are fundamental principles of physics that are divided into three laws of motion, each law focuses on each force of nature in physics, like the 1st law of motion focuses on inertia, the 2nd law of motion focuses on force, mass and acceleration, and last but not least, the 3rd law of motion focuses on action and reaction.

Let's start with Newton's first law of motion, the first law of motion focuses on inertia. What is inertia? Inertia is the resistance in which the object has a change in its state of motion. Newton's first law of motion states when an object stays at rest, tends to stay at rest, and when an object stays in motion, tends to stay in motion, when an external force acts on the object like gravity.

So, what does it mean by an object stays at rest, tends to stay at rest, and an object stays in motion, tends to stay in motion? Well, it means the object stays still when there is no force applied to it, like a soccer ball stays still, when a person doesn't kick it, and it also means the object stays in motion, when force is applied to it, like a person kicking a soccer ball, applying force.

The object's tendency to resist its changes in motion, is related to its mass. The more mass the object has, the more inertia has, and the less mass the object has, the less inertia has, (for example, a feather has lesser mass, it can move on its own, when applied by force like air). Newton's first law of motion and other laws of motion were named after English physicist and theologian himself, Sir Isaac Newton (1642-1727). Newton developed the laws of motion, including his first law of motion on his book which he published in 1687, titled in his original language Latin as "*Philosophiæ Naturalis Principia Mathematica*", which is translated to Modern English as "*Mathematical Principles of Natural Philosophy*".

You may also wonder what is the external force that is being applied on these objects in Newton's first law of motion? That external force in physics is called a net force, it is some unbalanced force that is originated in the object, and but cannot be cancelled out by another force. But if the net force is zero, it means that the force will be balanced in the object, and it will be cancelled out by another force. The magnitude will be equal, and the direction will be opposite. The object stays in motion, tends to stay in motion at the same speed and the same direction, unless an external force takes action towards the object.

So, does Newton's first law of motion have a connection with the Earth's revolution around the Sun in our Solar System? Well, yes, inertia plays a role in the Earth's revolution, because the Earth has the tendency to move in a straight line because of inertia itself. But the Earth is being controlled by the Sun's gravity to create an orbit. The Earth tends to move in a straight line, but the Sun's gravity pulls the Earth to move in an orbit. What does this all mean? Yes, it is confusing, but inertia's connection with Newton's first law of motion proves that the Earth's orbit results in a continuous "tug-of-war" situation as it goes back and forth, back and forth between the inertia force and the Sun's powerful gravitational force. So, eventually, the Earth's orbit always falls to the Sun's powerful gravity around it, rather than the tendency of moving to a straight line.

Here is the image of how inertia works in Newton's first law of motion by showing the Newton's cradle:

We move from the first law of motion to the Newton's second law of motion, which is a fundamental principle of motion developed by Sir Isaac Newton focusing on force, mass and acceleration. So, what is Newton's second law of motion, exactly?

Yes, I explained earlier, again, Newton's second law of motion, which is a fundamental principle of motion developed by Sir Isaac Newton, focusing on force, mass and acceleration. It also states that the acceleration of an object is directly proportional (meaning it is the same with force), when net force is acting on the object. For example: If you double the net force on the object acting on it, the acceleration will double as well, if the net force triples, the acceleration of the object will triple as well and so on.

Newton's second law of motion has a formula that describes the relationships of force, mass and acceleration, $F = ma$, which is the one of the most incredible and simple formulas in physics like $E = mc^2$. What does this formula actually mean? Well, $F = ma$ represents F as the net force applied on the object, m for mass of an object, and a for acceleration of an object.

You may also wonder, what about mass' involvement in force and acceleration's works in an object? Mass and acceleration's relationship is different from force and acceleration's relationship. Mass and acceleration are inversely proportional (meaning the mass is opposite with acceleration), For example: if you double the mass of an object while the net force is applied on the object, the acceleration will be halved, meaning the more the object is massive, the less the objects accelerates at an applied net force.

In physics and the world of the metric system, force is a unit of measurement as well, not just a natural thing in our universe, for example: $F = ma$ is converted in unit form as $N = kg \cdot m/s^2$. N for newtons, units that measure force, kg for kilograms, units that measure mass, and m/s^2, meters per second squared, not a unit, but derived units that divides distance by time squared.

Here is another burning question you may also ask while reading, does Newton's second law of motion have a direct and fundamental connection with the Earth's revolution around the Sun? According to the Newton's first law of motion, the Earth moves in a straight line because of inertia, being pulled back and forth between inertia and the Sun's gravitational force, causing the Earth to move on a curved orbit around the Sun, and the Sun's gravity is so strong that the Earth can revolve rather than move on a straight line.

But physicists and astronomers use advanced technology to calculate the exact acceleration, to observe the Earth's different speeds, slower when closer to the Sun, and faster when farther away from the Sun. Force is not just equivalent to mass multiplied by acceleration. Here is another formula not from Newton's second law of motion, but from Newton's Law of Universal Gravitation, $F = GmM/r^2$.

So, what does the $F = GmM/r^2$ formula represent exactly? Here is the bullet list below, to find the answers and the representations of the formula, yes, yes, you may also know that F is obviously force.

- F represents the gravitational force between the two objects, measured in Newtons (N).
- G represents the universal gravitational constant, it is a fixed value that is approximately 6.674 × 10^-11 N · m^2/kg^2 (force multiplied by distance squared divided by mass squared). It shows the strength of the gravitational field.

- M and m represents the masses measured in kilograms, the force is directly proportional to the two different products of their masses.
- R represents the distance between their centers of mass and their objects, measured in meters. The force is inversely proportional to the square of the distance, meaning that the distance increases, the force decreases rapidly or swiftly.

We are done with Newton's second law of motion, let's relocate and head on to Newton's third law of motion, for every action, there is an equal and opposite reaction. Newton's third law of motion has one of the most renowned, memorable and simple phrases in history and in the present around the world. You may also wonder, what is Newton's third law of motion, and what does this phrase mean, actually?

Newton's third law of motion, is a fundamental principle of motion that focuses on action and reaction forces, and it also states that for every action, there is an equal and opposite reaction. The phrase actually explains the fundamental properties of forces as interactions between two different objects. The first objects exerts an action force on the second object, then the second object simultaneously exerts a reaction force equal in magnitude and opposite in direction on the first object. What does it actually mean by forces as interactions between two different objects?

Well, force itself, isolated cannot exist in nature, but forces only arise and exist between the interactions of two different objects. So, you may also wonder, what does "equal in magnitude, and opposite in direction" mean? The reaction force is equal in magnitude, because the strength of the force exerted by the first object on the second object is exactly the same as the strength of the force exerted by the second object on the first object. The reaction force is also opposite in direction, because the two different forces act on the same line, but also act on different directions. Action and reaction forces do not cancel each other, because they act on different directions, for motion to take action, there must be a net force on a single object.

For example: a rocket lifts off, and flies to space, the rocket's engine burns fuel, and expels high-pressure and hot gas backward, and downward during liftoff, at a high velocity. The rocket exerts an action force on the gas. The expelled gases exerts an equal and opposite reaction force on the rocket itself, the force that makes the rocket go forward is called thrust. You may also wonder what is thrust? Thrust is an unbalanced net force, (the force explained on Newton's second law of motion), acting on the rocket, according to Newton's second law, this unbalanced net force causes the rocket to accelerate in the opposite direction from the exhaust.

One last question before this sub-chapter concludes: what connection does Newton's third law of motion have with the Earth's revolution around the Sun? SThe Sun exerts a gravitational force which pulls the Earth from orbiting around it. The reason behind the Earth's revolution, the Sun is more massive than the Earth, the Sun accelerates less than the Earth. The Earth accelerates continuously and orbits around the Sun. It states that the Sun exerts a reaction force equal in magnitude and opposite in direction on Earth, than the Earth itself exerts an action force on the Sun. Physics and Newton's laws of motion, have "exerted a force and pulled" our minds, and opened a world of wonder about how forces and objects react with each other. If you know what I mean.

CHAPTER 9.2

EVERYTHING YOU NEED TO KNOW ABOUT PHYSICS & KEPLER'S LAWS OF PLANETARY MOTION

HOW AND WHY DO PLANETS ORBIT AROUND THE SUN IN DIFFERENT ORBITAL PERIODS?

Have you ever wondered why do other planets in different orbital periods? What are Kepler's laws of planetary motion? Who developed this kind of laws of physics?

Kepler's laws of planetary motion, are a different kind of fundamental principle in physics that describes how planets orbit the Sun. The laws of planetary motion were developed by German astronomer and mathematician Johannes Kepler in two books "*Astronomica nova*" published in 1609 and "*Harmonices mundi*" published in 1619. What do Kepler's three laws of planetary motion state exactly and do these laws have a connection with Copernicus' heliocentric model?

Here are three laws of planetary motion, explained by German mathematician Johannes Kepler:

- Kepler's first law of planetary motion (published the book titled "*Astronomica nova*" in 1609) states that the planets orbit around the Sun at one focus, Each planet's orbit is not a normal circle, but an ellipse, wider than a circle. The distance between the Sun and the planet is not constant throughout the orbit.

- Kepler's second law of planetary motion (published the book titled "*Astronomica nova*" in 1609) states that the line connecting from the planet to the Sun sweeps out equal areas in equal times, meaning that when the planet orbits faster, it is closer to the Sun, and when the planet orbits slower, it is farther away from the Sun.
- Kepler's third and final law of planetary motion (published the book titled "*Harmonices mundi*" in 1619), the square of the planet's orbital period is directly proportional to the cube of the planet's semi-major axis. It means when the planet is farther away from the Sun orbiting, it takes a long duration of time, to complete one full orbit around the Sun.

In Kepler's third law of planetary motion, the orbital period, is an amount of time, the planet takes time to complete one full orbit around the Sun, (e. g. the Earth for 365.25 days). The semi-major axis, is the planet's distance from the Sun (e. g. the Earth has a distance of 15 × 10^6 km from the Sun).

So, what connection does Kepler's laws of planetary motion have with Copernicus' heliocentric model? Kepler's laws of planetary motion has a connection with heliocentrism, because these laws of planetary motion have the evidence of proving whether heliocentrism is true or not. Heliocentrism is the belief that Polish astronomer Nicolaus Copernicus conceptualized about in the early 1500s, Heliocentrism is a type of scientific ideology that states that the Sun is at the center of our solar system. The Sun being at the central or focal point of the solar system is the most crucial evidence that Kepler had proven on his laws of planetary motion, how planets orbit around the Sun, and why they orbit the Sun through orbital periods.

CHAPTER 9.3

PHYSICS & THE LAWS OF THERMODYNAMICS

DOES ENERGY RECREATE ITSELF AFTER BEING USED?

Have you ever wondered does energy recreate itself after being used inefficiently? What are the laws of thermodynamics in physics and are these fundamental laws crucial in everyday life?

The laws of thermodynamics are the fundamental principles in physics that focus on the relationships of heat, light, energy and mechanics. There are three laws of thermodynamics, I mean four, starting from the Zeroth Law of Thermodynamics to the Third Law of Thermodynamics.

The laws of thermodynamics were first developed by Rudolf Clausius and 1st Baron Lord Kelvin William Thomson, and Sadi Carnot (in his 1824 work "*Reflections to the Motive Power of Fire*" that explains the second law of thermodynamics and heat engines).

Let's start with the Zeroth Law of Thermodynamics, which explains about two systems are each with thermal equilibrium with a third system, they are both with thermal equilibrium with each other, meaning that the two systems are in contact with each other, and there is no heat flow between these systems, meaning that the third system has an equal temperature with the other two systems.

The Zeroth Law of Thermodynamics proves the transitive property of temperature, how it transitions from one system to another system.

Let's move from the Zeroth Law of Thermodynamics to the First Law of Thermodynamics. What is the first law of thermodynamics?

The First Law of Thermodynamics restates what the Law of Conservation of Energy states. Both the First Law of Thermodynamics and the Law of Conservation of Energy state that energy cannot be created nor destroyed, but it can change any form.

So, what does it mean by energy cannot be created nor destroyed, and can energy recreate itself after being used inefficiently? In the First Law of Thermodynamics and the Law of Conservation of Energy, energy cannot be created nor destroyed, but can change to any form, meaning energy can convert itself, from heat to light to chemical to kinetic as well.

For example: a light bulb is turned on, its filament produces heat energy then converts to light energy, then converts again to electrical energy. This is how energy works in our everyday life, it changes to any form, it isn't created nor destroyed.

Energy doesn't recreate itself after being used inefficiently, because energy inefficiency happens with the energy is less powerful and concentrated, it produces low-grade heat, and the cause of less energy.

The first law of thermodynamics is being proven by a mathematical formula of how energy changes from one form to another, $\Delta U = Q - W$. You may also wonder what is that triangular symbol on the formula? The triangular symbol used on formulas in physics and mathematics, it is called a delta (Δ).

The "$\Delta U = Q - W$" formula from the First Law of Thermodynamics, the first law is expressed as the change of internal energy is equal to the heat minus by the amount of work in the system.

- (ΔU): Internal Energy, the change in the system's internal energy
- (Q): Heat, the heat transferred in the system
- (W): Work, the work done on the system.

Let's head off and learn what the Second Law of Thermodynamics is. The Second Law of Thermodynamics states that in any reverse or spontaneous process, any entropy of a closed system increases, it means that any disorder in the universe is always increasing, and energy spontaneously flows from hotter to colder objects, like heat transfer does.

You may also wonder what is entropy, and a spontaneous process? An entropy is a measure of disorder or randomness in an energy system, while a spontaneous process is a process that happens on its own without any energy source.

While the entropy always increases, the universe is on a state of a greater disorder, when a solid ice cube melts into a puddle of water, it cannot freeze back again without an energy source like a freezer.

Why is the Second Law of Thermodynamics crucial anyway in our daily life? The Second Law of Thermodynamics is crucial in our daily life, because it explains the limitations and minimum life of these energy-efficient machines, they may lose energy, as time goes forward, it is entropy, disordering energy.

Engineers understand the Second Law of Thermodynamics, and use it to build more super-efficient electrical devices that may not or may lose energy in the future.

For example: In South Africa, Eskom, the electrical state-owned company that generates and transmits approximately 90% of the country, is experiencing load-shedding problems every year since 2007, and this might be a good example of what the Second Law of Thermodynamics explains about the loss of energy and the arrow of time.

Let's move from the Second Law of Thermodynamics to the Third Law of Thermodynamics, so, what exactly does the Third Law of Thermodynamics focus on? Well, the Third Law of Thermodynamics states that the perfect crystal can reach an absolute zero (0 Kelvin) with no entropy.

But it is unattainable, because it is impossible to reach absolute zero, it requires infinite number of steps in order to remove all the thermal energy in a system.

So, why are these laws of thermodynamics crucial in our daily life? The laws of thermodynamics play an important role, because they manage energy transfers (from hot to cold), and fundamentally knowing our universe's workings with energy.

Starting from the First Law of Thermodynamics, it states that energy cannot be created nor destroyed, but can be changed to any form. For example: Us, humans, we eat food, the chemical energy stored in our food we eat is converted into kinetic energy (energy that produces motion), and the thermal energy (energy that produces heat). The energy used to power a light bulb doesn't vanish, but electrical energy is converted to heat and light energy. The petrol's chemical energy is released through combustion (burning of fuel), and converts to mechanical energy which moves the car.

The Second Law of Thermodynamics, states that in any spontaneous process, the total entropy of a closed system increases. Our human bodies produce heat, and releases it, it develops entropy (disorder of energy). Heat used on coal-powered stations is being released, wasting energy, and causing load-shedding and disrupting power cuts. In the Second Law, energy is never 100% efficient, but it can be wasted when heat lacks, and entropy increasing. The Third Law of Thermodynamics, states that the perfect crystal reaches at 0 Kelvin or absolute zero, but it has no entropy, and it is unattainable to reach the absolute zero, because it requires infinite number of steps to remove heat.

How the Third Law interacts with daily life, it helps us understand why we use refrigerators and freezers to preserve food, and freeze it to prevent spoilage. Physics and the laws of thermodynamics, have taught us how and why energy is so important, and changes or converts in any form,

CHAPTER 10

EVERYTHING YOU NEED TO KNOW ABOUT PHYSICS & THE FUTURE

Have you ever wondered how physics and the future align with each other? You may also wonder how will the future look like, and will there really be an extinction of humans in 2050? Well, we will never know, until we warp 20 more years forward to the future.

In this final chapter of the book, we will focus on topics that dive deep on how physics connects with the future:

- The Future of Gaming: How advanced technology like virtual reality, augmented reality, and mixed reality will shape the future of video gaming and physics engines forever.
- The Future of Transportation: How advanced transportation technology like hyperloop and maglev can help a passenger reach to his workplace early at a second or a microsecond maybe, it depends on the future.
- The Future of Time: Is time going to be more faster in the future depending on the Earth's revolution speed.

Imagine what the world will be with the help of advanced physics, technology, science and any field of study. As dreamers who believe to become unstoppable like to say, the future starts now, and physics' connection with the near and far future, begins to glow with the never-ending curiosity of wanting to know more about the future of physics. Let's start now.

CHAPTER 10.1

PHYSICS & THE FUTURE OF GAMING

WILL VR, AR, AND MR REALLY HELP US INTERACT WITH THE REAL-WORLD MIXED WITH VIRTUALITY WITH THE HELP OF PHYSICS ENGINES?

Have you ever wondered how will video games look like in the future, and will there be a PlayStation 1000? Well, not exactly, but video games will look super-realistic like us, humans in real life. These upcoming futuristic video games will have super-advanced graphics and the physics engines will improve with the video game characters' interactions with the game, looking more sleek and realistic.

So, the future of gaming starts now, in the present, Video gaming companies, like Meta (responsible for Facebook, Instagram and WhatsApp), developed a virtual reality headset known as the Meta Quest. The development for the incredible virtual reality headset started in 2018, for R&D (research and development), then the VR headset released in 2019. But what is virtual reality's connection with physics and the future of gaming?

Virtual reality's connection with physics, is actually the reason that VR systems use physics to simulate the character's body movements and its interactions with the real world. Physics engines like NVIDIA PhysX, simulate the video game's realistic physics like falling down, or flying and collision physics, colliding and interacting with other objects in the game.

So, how does virtual reality work? Virtual reality is an advanced technology developed by scientists and it works by using a VR headset with two high-res screens with special lenses to present different images on each eye, creating a stereoscopic 3D effect, that gives the illusion of depth, that looks like you are actually in the video game, like literally in the video game.

Sensors in the headset, help you track your head and hand movements in the game. The combination of the visual, auditory, and haptic touch senses, trick your brain into looking like you are actually in the video game, and immersed in the computer-generated environment. So, you may wonder that augmented reality and virtual reality are the same. But, you are actually wrong, augmented reality (AR) is a technology that creates 3D computer-generated digital images like monsters from a known video game in the real world, at the exact time. While virtual reality (VR) on the other hand, is an advanced technology that simulates a 3D computer-generated environment inside your headset, to look like you are actually present in the video game, with the help of sensors.

Here is the image below behind the difference between augmented reality and virtual reality, image courtesy: Freepik:

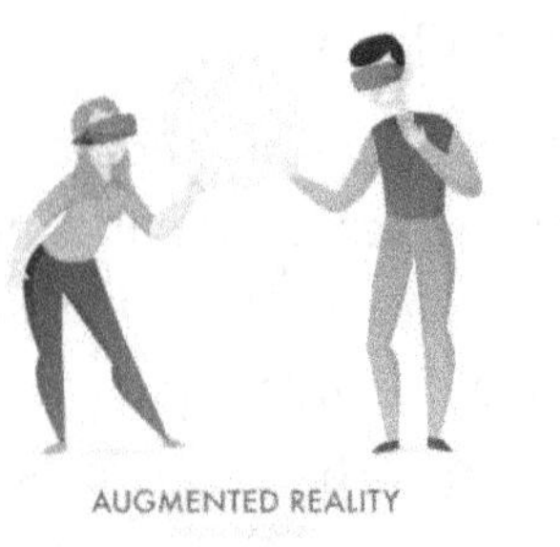

What kind of technology is augmented reality, and why is it so different from virtual reality? And why is it so interesting about augmented reality, anyway? It is just a technology that creates computer-generated digital images into the real world using smartphones and even augmented reality headsets.

Augmented reality (AR) is an incredibly astounding kind of technology that creates computer-generated images in the real world, from your smartphone or tablet cameras. Augmented reality is interesting because of its simulation of these computer-generated images in the real world, looking like those images are actually there but when we see on our smartphone or tablet displays, seeing an illusion.

So, you may also wonder how does this augmented reality technology work? Here is the list below:

- Sensors and cameras: devices like smartphones and tablets use cameras to capture the surroundings of our real-world environment and sensors like accelerometers, gyroscopes and GPS are used to determine the device's location, orientation and the objects' position in three-dimensional (3D) space.
- Processing Units: Powerful device processing units like CPU (central processing unit) or GPU (graphics processing unit) are used to render graphics of the computer-generated image, and analyze the video feed of the image in real time or showing stats of the image, from the processor's performance.
- Display Devices: The display devices are crucial in augmented reality technology because they are obviously helping the user to see the merged reality using

smartphone or tablet screen or any augmented reality headset like a Microsoft HoloLens headset.

So, what is the “magic” thing that happens in both virtual reality and augmented reality? The actual “magic” thing that happens in both virtual reality and augmented reality leads to the creation of some psychological phenomenon in your mind, to believe that you are actually present in the virtual and augmented reality simulation, the sense of presence.

In your mind, in virtual reality, the sense of presence you have inside the simulation, completely immerses the user to a real computer-generated environment, feeling like you are actually there in the environment, looking like you are there and forgotten you wore a headset, but it doesn’t like you wore a headset in the simulated environment, but also looks like you have entered in a different, amazing and a realistic world.

Then, in your mind, in augmented reality, the sense of presence you have with the simulation on your AR headset and the real world, the digital images is being overlaid or created onto the real world, showing that as if those digital images are actually there, allowing the user to interact with those virtual elements that those digital images are creating, as if they are a natural part of their physical environment.

Here is another question, what link does VR and AR technology have with the near and far future? The link that these two different kind of reality technologies, will have in the near and far future? Let’s start with the near future, so, virtual reality and augmented reality will both dominate video game consoles like PlayStation and other consoles like Xbox Series, and in the far future, both VR and AR technologies will advance into Mixed Reality (MR) and some kind of advanced reality technology that will be known as Extended Reality (XR), but was initially developed in the 1960s, starting the XR timeline. You will learn and clarify more about mixed reality and extended reality, later on. So, what is the physics behind these technologies in both the near and the far future?

The actual physics behind virtual reality and augmented reality in the near future:

- The near-future technology that will advance virtual reality, starts with how optics and lenses will work in the near-future, both virtual and augmented reality headsets use displays that shows different images on each eye, mimicking or copying binocular disparity, the brain uses to perceive depth into the real world.
- What is binocular disparity? Binocular disparity is the difference in the image of your world, where your oculus sinister (left eye) against your oculus dexter (right eye), caused by the horizontal separation in which the brain uses a 3D perception of depth to see clearly, exactly what VR and AR technology's future advancements will be.
- And in sensor and tracking technology, sensor and tracking devices used on both VR and AR headsets will improve, like gyroscopes, accelerometers, and magnetometers, that will rely on the principles of motion and forces, to track both head and hand movements in real time.
- And then the physics behind virtual and augmented reality in the far-future, focuses on how VR and AR technology will have a link with brain-computer interface technology, some type of technology developed by Neuralink (Elon

> Musk's neurotechnology company). What is brain-computer interface technology? Well, in the far-future, BCI will allow users who are inside the VR or AR computer-generated simulated environment, to control and interact with the virtual world using their thoughts, from the electrochemical signals of the brain itself.

Yes, both reality technologies, VR, AR, and MR will help us interact with the real world, but with different virtual elements with the help of physics engines and advanced technology like brain-computer interfaces. Physics engines used for simulation and game developments to calculate the character's animation and interactions, and everything physics like colliding with other objects, gravity and so on. But you haven't learnt about what mixed reality and extended reality are exactly? Let's find the answers to know what these two technologies are. Starting with mixed reality, what is mixed reality or MR for short? Mixed reality (MR) is an advanced technology that merges both virtual reality and augmented reality, leading to users can see and interact virtual objects in the real world. Mixed reality see-through headsets like Apple Vision Pro, with cameras and sensors to map surroundings, enabling real-time interactions, between real people, objects and digital stuff.

On the other hand, what is extended reality (XR)? Extended reality is an advanced technology that merges all virtual reality, augmented reality and mixed reality. XR technology uses VR to interact with virtual objects in a virtual environment), AR to see digital images interact with the real world, and MR as the combination of actual and virtual objects in real time. Physics has taught us how VR, AR, and MR technologies will impact the future of gaming, by allowing the user to interact with different worlds and also different universes, but in the real world.

CHAPTER 10.2

PHYSICS & MAGLEV

IS MAGLEV PHYSICS GOING TO MAKE ALL MODES OF TRANSPORTATION LEVITATION-POSSIBLE?

Have you ever wondered how will the future look like with cars now hovering in the air, not drifting on the ground? Well, in this sub-chapter, you will learn how magnetic levitation physics will interact with the future of transportation.

Firstly, what is magnetic levitation, and why is it involved in transportation and its future? Magnetic levitation or maglev for short, is a type of high-speed transportation system using magnets to levitate a train above the railway, preventing wheel friction, for a smoother, faster and a more silent ride when commuting from home to work, and from work to home.

How does maglev transportation work? Maglev works with electromagnets above the railway and under the train, providing both lift and propulsion along the railway track. By the train floating inches above the track, the maglev train travels at a rate of over 300 to 600 miles per hour (mph).

Here is the image at the bottom right side of the page, showing China's maglev powered train, the Shanghai Transrapid.

Maglev actually works, when the powerful electromagnets on the train interacts with the coils on the railway track, they either attracting which leads to EMS (Electromagnetic Suspension) or repelling which leads to EDS (Electrodynamic Suspension) the train, lifting it off the track. The propulsion in the train works, when the magnetic fields change in the railway track, pushing or pulling the train forward. The railway track is structured as a U-shaped concrete structure containing magnetic coils.

What is so fascinating about maglev trains? The reason behind the fascination of maglev trains, is because they prevent wheel friction, no screeches during the ride, just a smooth, relaxing, quieter and an extremely faster ride, with no train interruptions. The maglev trains are fascinating because they are weather resistant, less affected by rain, snow and ice and other extreme weather conditions, than normal trains.

So, the sub-chapter's question, is maglev physics going to make all modes of transportation levitation-possible? Is there a possibility that other types of vehicles are able to levitate in the future?

It is actually impossible for all other types of vehicles to levitate in the future, like airplanes and private jets, but Hyperloop technology is the solution to improve transportation with maglev physics.

What is hyperloop technology? Hyperloop technology is a different type of transportation technology that builds a vision of high-speed vehicles where they travel on clear tubes that help them accelerate faster with the help of magnetic levitation where the rails are filled with powerful electromagnets called neodymium magnets. Well, you may also wonder what is EMS and EDS and what are neodymium magnets, exactly?

EMS stands for Electromagnetic Suspension uses magnetic attraction between electromagnets on a vehicle and the railway track or road to levitate or support the weight of the vehicle, giving a contactless, super-incredible and super-smooth ride to prevent wheel friction and by actively controlling the magnetic forces to counteract bumps.

While EDS stands for Electrodynamic Suspension is a maglev system that uses the repulsive force between the powerful electromagnets and the induced electric currents in a conductive railway track to lift and stabilize the vehicle, creating a gap preventing physical contact between the vehicle and the railway track.

Then, neodymium magnets, on the other hand, are powerful electromagnets that they are composed of three chemical elements, neodymium (Nd), boron (B) and iron (Fe). Why are neodymium magnets so strong, in the railway? In the microscopic inner working of a neodymium magnet, the neodymium atoms act like powerful magnetic dipoles (a pair of magnetic poles that are separated over a distance), and iron adding strength, and boron stabilizing the structure of the magnet.

Here is the last question before this sub-chapter ends, is maglev physics the future of all modes of transportation? Yes, and no, the advantage side is that maglev physics is only required for the future of ground transportation like buses, cars and vans, and then the disadvantage side is that maglev physics isn't required for the future of air and sea transportation. Physics and the world of magnetic levitation, have taught us how they work together to lift a train using EMS and EDS, how magnets repel and attract, and also how maglev physics is crucial for the future of ground transportation.

CHAPTER 10.3

PHYSICS & THE FUTURE OF TRANSPORTATION

ARE SCIENTISTS AND TECH PIONEERS GOING TO DEVELOP NEW DEVICES AND SPACESHIPS TO TRAVEL THROUGH SPACE?

Have you ever wondered how transportation will end up being like, in the future? It actually means, that how the future of transportation will look like more than air, sea and land transport, but beyond like space transportation. Well, in this sub-chapter, we focus on how advanced transportation will look like in the future, and the question is, are we humans going to develop new devices and spaceships to travel through space? There are many predictions of how transportation and how the world will look like in the near and far future, like cyberbuildings and spaceships in the air, looking so futuristic, the ones you see on science fiction movies and series.

So, let’s start from how transportation will develop in the near future. Transportation in the near future will develop, starting with electric vehicles like, for example, a Tesla Cybertruck model, which support smart charging systems, and its support with renewable energy sources like photovoltaic energy will be crucial. Autonomous vehicles like self-driving cars are also involved in the near future of transportation, and its technology will be highly advanced.

The near future of transportation might involve space tourism, in air and space transportation, delivery high-tech drones and electric vertical take-off and landing aircraft or e-VTOL for short, e. g. flying taxis will improve transportation and will be tested and be used on fast deliveries and short urban trips. While the far future of transportation, on the other hand, will be more futuristic than the near future of transportation.

Advanced transportation technologies like the Hyperloop technology will be involved, the Hyperloop technology will involve pods traveling in a low-pressure and see-through tube, at extremely high speeds, that could link and help you travel through different cities, replacing air transportation. Hypersonic flight technology will also be involved, traveling faster than the speed of sound in dry air which is 3.43 m/s, that reduces long-distance travel duration, helping you travel from Johannesburg to London, in two hours, or wherever you are in the world to other parts of the world.

It is affirmative and true that scientists and technologists are finding new ways to develop new transportation devices and spaceships to help you travel through space, and on Earth, at a second, or even short duration of hours. Beyond space tourism, space transportation systems like spaceships and space elevators could become easier, moving people and supplies in orbit and supporting out-of-this-world human settlements. What is the actual physics behind the near and far future of transportation? The transition from normal vehicles to electric vehicles will be much more easier, because it relies on advances supported by electromagnetism, material science and energy storage. Physics and its connection with the future of transportation have taught us, how they will work together perfectly for the environment, reducing pollution from advanced material science, and maglev transportation modes as well. May time go forward and see how transportation will improve, and more highly advanced.

CHAPTER 10.4

PHYSICS & TIME II

IS TIME GOING TO BE MORE FASTER IN THE FUTURE?

Have you ever wondered why time is going so fast? It means, that it was 5 years ago, when you were ten years old, embracing some incredible and some heart-breaking moments. In this final sub-chapter of the book, we will uncover the secrets of time, and how will it be in the future, will it go slower or faster, or following some pattern that we humans don't even know. It all started from the biggest cataclysmic explosions of the Big Bang, to the early Egyptian and Babylonian measurements of time in the BCE era, to the modern crucial use and management of time.

Some people are asking what is time, who created time, and who discovered time? Time is the most fundamental aspect in our universe, measured in hours, minutes, seconds, milliseconds, microseconds and so on. No one discovered time, but in belief, God created time since he created the universe on the Genesis book in the Bible. The Egyptians and Babylonians are the only ancient civilizations that discovered time, by measuring it using sundials. *Discover what sundials are on page 97.*

What is the actual physics behind time going faster in the future? The actual physics behind time going faster, is all comes down to Einstein's general theory of relativity, focusing on three spatial dimensions and one time dimension. But what link does Einstein's general theory have with the future?

Gravitational time dilation is a type of phenomenon in physics that when gravity is stronger, time ticks slower, and when gravity is weaker, time ticks faster, like clocks on Earth ticks slower than the clocks on satellites being pulled on the Earth's gravitational pull. The reason behind the link of Einstein's general theory of relativity in the future, is time dilation, black hole physics and a possibility of reaching the speed of light.

The theory predicts that the traveling at speeds close to the speed of light, would cause time to be slower than time on Earth, it might be some kind of theoretical possibility to reach the speed of light and time travel to the future. In the future, time dilation might be involved in the future of space exploration and travel.

Here is another question to conclude this sub-chapter, is time travel going to exist in the future? Time travel already exists, and it will be possible in the future, but traveling back to the past is likely impossible because it requires infinite energy to reverse entropy (disorder of energy). Physics and time have a mind-boggling and flabbergasting connection, because time is the most amazing thing in the universe, because it is not just a measurement, a fourth dimension according to Einstein's general theory of relativity, how it has a link with three spatial dimensions (length, width and height). The final sub-chapter concludes, well you had an incredible journey from how physics is involved in gaming, and world we live in to how physics is involved with the near and far future and how physics interacts with everything.

Bibliography

The following websites and online articles were used to support the research, explanations, and information presented in this book. These sources helped verify facts, expand knowledge, and provide accurate technical and crucial details. Listing these sources helps maintain accuracy and proper credit to the original information.

- SDLC Corp: “How Physics is Important in Video Gaming and Its Connection with Gaming?”, *https://sdlccorp.com/post/exploring-the-importance-of-game-physics-in-console-games/*
- Wikipedia: “Physics’ Importance with Game Development and SDKs”, *https://en.wikipedia.org/wiki/Video_game_development*
- Cosmos at your Doorstep: “What is the spacetime curvature in Einstein’s general theory of relativity?”, *https://cosmosatyourdoorstep.com/2017/09/28/einstein-space-time-curvature/*
- Wikipedia: “How Does An Incandescent Light Bulb (Edison’s Invention) Work?”, *https://en.wikipedia.org/wiki/Incandescent_light_bulb*
- Wikipedia: “How Does Light Work?”, *https://en.wikipedia.org/wiki/Light*
- Wikipedia: “Electromagnetic Radiation, How Does It Work in the Electromagnetic Spectrum?”, *https://en.wikipedia.org/wiki/Electromagnetic_spectrum*

- Department of Energy: "Our Expanding Universe, Dark Matter and Dark Energy and the Big Bang, *https://www.energy.gov/science/articles/our-expanding-universe-delving-dark-energy*
- Wikipedia: "Understanding the Fourth-Dimensional Object, the Tesseract and More about Mathematical Dimensions", *https://en.wikipedia.org/wiki/Tesseract*
- NASA Glenn Research Center: "Scalar and Vector Quantities: Magnitude and Direction", *https://www.grc.nasa.gov/www/k-12/airplane/vectors.html*
- Study.com: "Metric Vs. Imperial System of Measurement and the History of These Systems", *https://study.com/academy/lesson/metric-system-vs-imperial-system.html*
- Study.com: "Speed and its Involvement in Physics", *https://study.com/academy/lesson/measuring-the-speed-of-an-object-physics-lab.html*
- Unrevealed Files: "How Digital Data Interacts with Physics", *https://www.unrevealedfiles.com/digital-physics-exploring-the-universe-as-a-computational-system/*
- Innovation World: "Dynamite's Invention Replaces the Dangerous, Unstabilized Nitroglycerin", *https://innovation.world/invention/dynamite/*
- Wikipedia: "The Famous Kite Experiment, and the Test whether Lightning is Electrical or Not, by Benjamin Franklin, *https://en.wikipedia.org/wiki/Kite_experiment*

About the Author

Bongani Trinity Banda, is an author, future physicist, who has dream of becoming a future physicist, and other careers that he wants to be, like a radio presenter and an author as well. He has a strong inspiration in physics, broadcasting and writing non-fiction books like this book. Bongani has an infinite number of questions to ask how our world works, and beyond more.

He has strong inspirations from famous physicists, Brian Cox, Hakeem Oluseyi, Neil DeGrasse Tyson, Jim Al-Khalili, David Spiegelhalter, and wants to be like them, be informative to people, and loves exploring more, he is also a gamer, developing critical thinking skills from puzzle video games like Portal 2, a physics based video game released in 2011 on consoles.

Bongani is also enthusiastic, when it comes to discovering something he doesn't know and wants to know more about it, opening a world of curiosity in his mind. He watches informative documentaries and reality shows on Discovery Channel, and other factual channels, in which learning is his kind of entertainment.

"Dream big, keep shining, I hope this book opened up your curiosity, and answered to your previous unanswered questions about how physics is involved and how is it crucial in our daily life, Thank you, readers, for reading my book. I am appreciating it, and keep being curious, because curiosity is the key to finding success, also education."

– Bongani Banda.

www.ingramcontent.com/pod-product-compliance
Lightning Source LLC
LaVergne TN
LVHW052259200926
848481LV00002B/412